GOSS'S ROOFING READY RECKONER

GOSS'S ROOFING READY RECKONER

FROM TIMBERWORK TO TILES

INCLUDING METRIC CUTTING AND SIZING TABLES
FOR TIMBER ROOF MEMBERS

Fifth edition
C. N. Mindham

WILEY Blackwell

This edition first published 2016

© 1948, 1987, 2001, 2008 the estate of Ralph Goss, Blackwell Publishing and Chris N. Mindham
© 2016 by Chris N. Mindham

Registered office
John Wiley & Sons, Ltd, The Atrium, Southern Gate, Chichester, West Sussex, PO19 8SQ, United Kingdom.

Editorial offices:
9600 Garsington Road, Oxford, OX4 2DQ, United Kingdom.
The Atrium, Southern Gate, Chichester, West Sussex, PO19 8SQ, United Kingdom.

For details of our global editorial offices, for customer services and for information about how to apply for permission to reuse the copyright material in this book please see our website at www.wiley.com/wiley-blackwell.

Library of Congress Cataloging-in-Publication Data

Names: Mindham, C. N. (Chris N.), author. | Goss, Ralph. Roofing ready reckoner.
Title: Goss's roofing ready reckoner : from timberwork to tiles.
Other titles: Roofing ready reckoner
Description: 5th edition / C.N. Mindham. | Oxford, United Kingdom : John Wiley & Sons, Inc., 2016. | Includes bibliographical references and index.
Identifiers: LCCN 2015044551 (print) | LCCN 2015045710 (ebook) | ISBN 9781119077640 (pbk.) | ISBN 9781119077664 (pdf) | ISBN 9781119077657 (epub)
Subjects: LCSH: Roofs–Handbooks, manuals, etc. | Carpentry–Mathematics–Handbooks, manuals, etc. | Roofing–Handbooks, manuals, etc. | Engineering mathematics–Formulae–Handbooks, manuals, etc.
Classification: LCC TH2401 .G67 2016 (print) | LCC TH2401 (ebook) | DDC 694/.2–dc23
LC record available at http://lccn.loc.gov/2015044551

A catalogue record for this book is available from the British Library.

Wiley also publishes its books in a variety of electronic formats. Some content that appears in print may not be available in electronic books.

Cover illustration by kind permission of Kier Living Ltd.

Set in 9.5/11pt Universe by Aptara Inc., New Delhi, India

Printed in Singapore

M000122_051121

CONTENTS

Contents

Contents

1 INTRODUCTION AND ACKNOWLEDGEMENTS

The aim of this book, when first published in 1948, was to provide quick reference tables for the length and angles of cut for timber members in a traditional cut roof construction. Today, when many houses use trussed rafters for their roof construction, there is still a need for some parts of those roofs to be built using traditional methods, especially with the ever-increasing use of attic roof structures. The renovation of older roofs, extensions and conversions all require knowledge of roofing from wall plate to ridge, and the correct detailing of the roof covering materials themselves.

Relaxation in planning controls has allowed a wide range of smaller buildings to be constructed without planning permission. These include sheds, garages, garden office buildings and workshops. New information on the limitations of building profile and construction is included in this edition, together with some helpful drawings.

This book assumes that a basic architectural design of the roof to be constructed is already completed, that is, the span, pitch, length and any additional supporting walls. Guidance is given on how to calculate the size of individual roof member timbers, the cutting length, the angles and the compound cuts. The tables shown are based on BS5268 'Structural use of timber' and a comparison of timber sizes using Euro code 5 'Design of Timber Structures' is shown as both

Goss's Roofing Ready Reckoner: From Timberwork to Tiles, Fifth Edition. C. N. Mindham.
© 2016 John Wiley & Sons, Ltd. Published 2016 by John Wiley & Sons, Ltd.

design documents are currently acceptable by building control regulations. The book now also includes all aspects to be considered when choosing the roof covering, including the suitability of the tiles or slates for the pitch and exposure of the roof concerned, the choice of a 'warm' or 'cold' roof, the considerations to be given to the correct insulation, and the possibilities and avoidance of condensation within the roof space by dealing correctly with ventilation.

Solar panels, now frequently fitted to both new and existing roofs, impose different loads on roof structures and these have been addressed in a new chapter.

Finally, Health and Safety matters are addressed, including the 'Working at Heights Regulations', loading the roof structure with the roof coverings, lifting components, and the correct use of preservative-treated timber.

ACKNOWLEDGEMENTS

My thanks are due to all the manufacturers who have allowed me to use their product illustrations. I am particularly indebted to Robinson Manufacturing Ltd, for the help they have given on engineering timber components, and especially for help on spandrel panels. Thanks are also due to Kier Living Ltd for access to their construction site for photography for the cover of this book.

The Trussed Rafter Association helped with solar panel fixing research, and Anthony Gwynne kindly contributed with his overview of the building regulation requirements with regards to thermal performance.

2 ROOFING TERMINOLOGY

The main terminology used for roofing is listed below (see Figures 2.1–2.4).

Wall plate   The 'foundation' of the roof, usually 50 × 100 mm wide, must be bedded solid, level and straight on the top of the wall, or nailed to the timber-framed panel and strapped in place to prevent movement from the structure.

Purlin   The member carrying part load of the long common rafters, traditionally placed at right-angles to the rafter but now more commonly fixed vertically.

Pitch   The angle made by the slope of the roof with the horizontal. This may be stated in degrees on the drawing, it may have to be measured by protractor from the drawing, or it may have to be calculated by measurement if the new work is to match an existing roof.

Ridge   The timber at the top of the roof where the rafters meet, giving a longitudinal tie to the roof structure, commonly 38 mm thick, and of a depth equal to the top cut on the rafter plus approximately 38 mm. This depth will depend on the pitch of the roof and the tile batten thickness.

Common rafter   The timber running from the ridge, down over the purlin if fitted, over the wall plate, and to the back of the fascia.

Goss's Roofing Ready Reckoner: From Timberwork to Tiles, Fifth Edition. C. N. Mindham.
© 2016 John Wiley & Sons, Ltd. Published 2016 by John Wiley & Sons, Ltd.

Jack rafter The timber running from the hip rafter down over the purlin if fitted, over the wall plate, and to the back of the fascia.

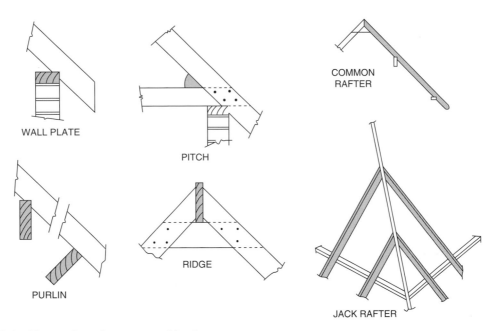

WALL PLATE

PITCH

COMMON RAFTER

PURLIN

RIDGE

JACK RAFTER

Figure 2.1 Illustration of terms used in chapter.

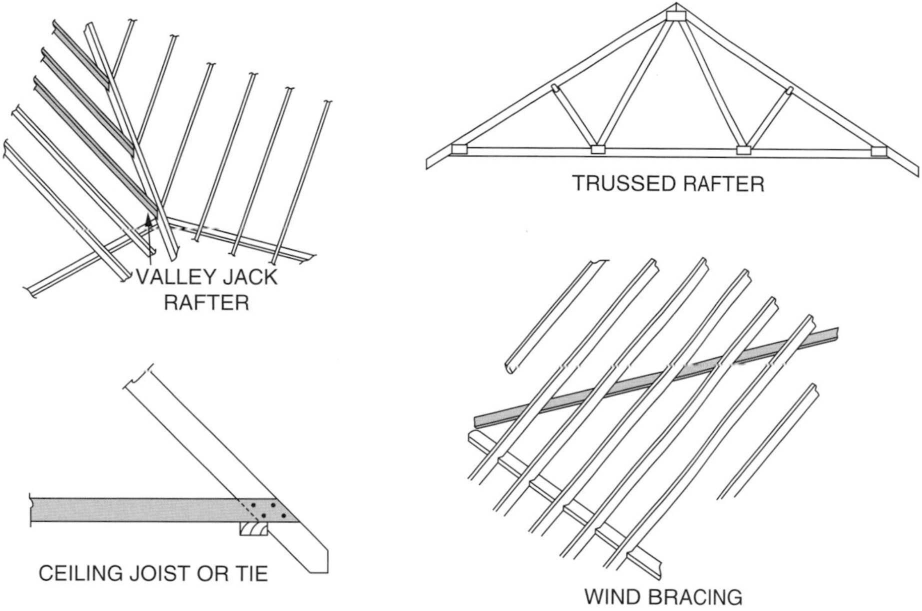

Roofing terminology

VALLEY JACK
RAFTER

TRUSSED RAFTER

CEILING JOIST OR TIE

WIND BRACING

Figure 2.2 Illustration of terms used in chapter.

Valley jack rafter   The timber running from the ridge, down over the purlin, down to the valley board or rafter.

Trussed rafter   A prefabricated framework incorporating rafter, ceiling joist (or tie), and strengthening webs forming a fully triangulated structural element.

Ceiling joist or tie   Timber supporting the ceiling of the building, but often importantly 'tying' the feet of the common and jack rafters together thus triangulating and stabilising the roof.

Wind bracing   Usually 25 mm × 100 mm timber nailed to the underside of rafters and trussed rafters running at approximately 45° to them, to triangulate and stabilise the roof in its vertical plane.

Attic or room-in-the-roof truss or trussed rafter   This popular truss shape allows the use of steeper pitch roof voids for habitable accommodation. There are no set minimums for dimensions H and W, but 2.3 m and 1.2–1.5 m are practical recommended minimums unless the room is to be restricted to storage only (see Fig. 2.3)

Longitudinal bracing   Usually 25 mm × 100 mm timber nailed to the underside of rafters and trussed rafters both at the ridge position on a trussed rafter roof, and at ceiling joist level on all roofs, to maintain accurate spacing and stiffening of the members to which it is fixed (Fig. 2.4).

Hip or hip rafter   This is a substantial timber member running from the corner of the roof at wall plate level to the end of the ridge. In some designs the hip may stop lower down the roof, producing a small gable at high level (Fig. 2.4).

Roofing terminology

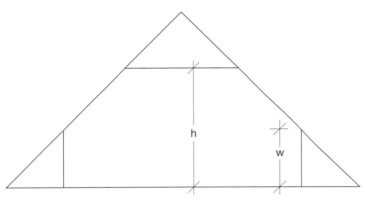

Figure 2.3 Attic or room-in the-roof truss.

Birds mouth   The cut in rafters at the fixing point to the wall plate and/or the purlin (where purlins are fixed vertically); this should leave at least 0.7 × the depth of the rafter to give the strength necessary for the rafter to continue to provide an overhang to the roof. If a common rafter is fitted as part of a trussed rafter roofing system, then the 0.7 × the depth of the rafter must be the same as the depth of the rafter on the trussed rafter component. (Fig. 2.4)

Fascia   Board fixed to the rafter feet, supporting both gutter and soffit (Fig. 2.4).

Soffit   Timber board or sheet material used to close off the overhang between the back of the fascia and the wall. This soffit may have a roof ventilation system built into it (Fig. 2.4).

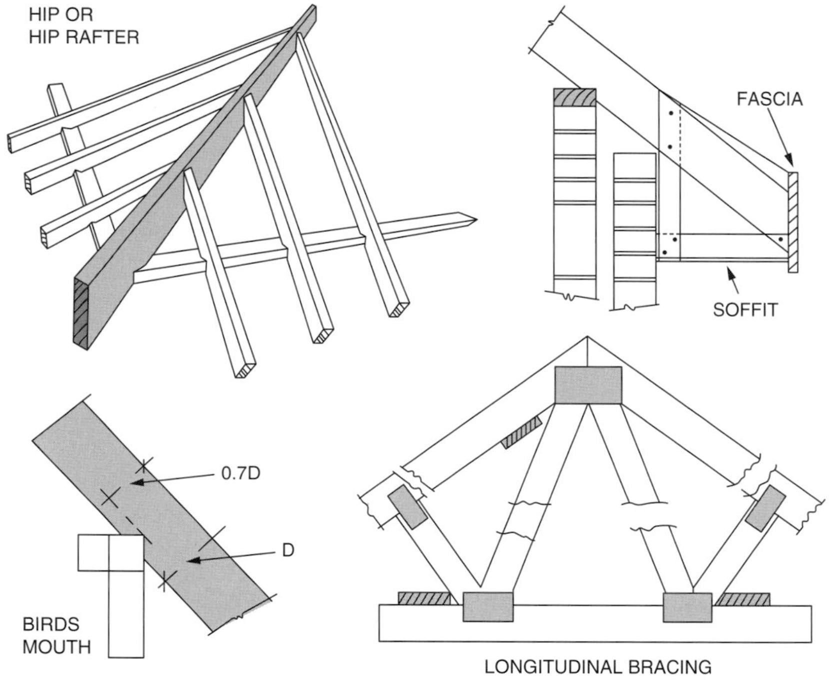

Figure 2.4 Illustration of terms used in chapter.

HIP OR
HIP RAFTER

FASCIA

SOFFIT

0.7D

D

BIRDS
MOUTH

LONGITUDINAL BRACING

3 CALCULATING THE SIZE OF TIMBER MEMBERS

Knowing the overall dimensions of the roof, that is, the span over the wall plates, pitch of the rafters, length between the gables or hips, any internal supporting walls and the specification of the roof covering, the following data will help to design the size and strength specification of the individual roof members themselves.

Roof structure design must satisfy the requirements of the building regulations for the individual countries of the UK. Currently (at the time of writing this edition), the building regulations can be satisfied by designing either with the British Standard BS 5268-2 2002 'Code of Practice for Structural Timber Design. Permissible stress design materials and workmanship' or BSEN 1995-1-1 'Eurocode 5 Design of timber structures'. Common rules and rules for building, the UK National Annex to EC5 and Published Document PD6693-1 'Recommendations for the design of timber structures to Eurocode 5', as above. Clearly, at some time in the future the building regulations will be amended to allow only the EC5 design standards. A limbo situation exists in the timber design industry at present, with some trussed rafters being designed on BS5268-2 and some on EC5. For individual timber roofing members, Trada have two sets of span tables available, one on each of the design methods above. These cover rafters, purlins, ceiling joists and ceiling binders,

Goss's Roofing Ready Reckoner: From Timberwork to Tiles, Fifth Edition. C. N. Mindham.
© 2016 John Wiley & Sons, Ltd. Published 2016 by John Wiley & Sons, Ltd.

plus floor joists and trimmers as well as flat roof joists. Designs based on either of these design documents should satisfy local building control. The Trada Technology Ltd documents are:-

- SPAN TABLES for solid timber members in floors, ceilings and roofs of dwellings, 2nd edition.
- Eurocode 5 SPAN TABLES for solid timber members in floors, ceilings and roofs for dwellings.

The first document is based on BS5268-2 and the second on EC5 (see Bibliography for full contact details to obtain the publications). The term 'solid timber' describes natural sawn timber from logs, and not engineered timber components such as Ply Box beams, or 'I' beams.

A brief word on timber size design considerations. The loads to be supported by the roof structure are made up of a number of elements:

(a) The roof covering: tiles, slates, etc.
(b) The self weight of the structure: timber, felt, battens, insulation, and ceiling if an attic structure, plus water tanks as necessary.
(c) Snow load.
(d) Wind load.

Referring to the above:

(a) is a statistic relating to the roof covering type and can be obtained from the product manufacturer.
(b) is calculated from the known weight of individual members.
(c) is a variable with the roof shape and pitch and height above ground and sea level, and is also a variable depending on the location of the building geographically within the UK.
(d) this is also a variable based on geographically variable information and is also affected by exposure, that is, the altitude of the building and its proximity to coastal or other areas of high wind exposure.

For the above reasons, any calculation for designing the roof member size, must take all of the above into consideration. Thus, most standard timber sizing data has geographical limitations to its use. The tables reproduced in this book have been designed to illustrate the process to be used when sizing timber members in the following worked example. The Trada tables referred to above are more comprehensive in terms of different loadings, different spacings, and give data for both C16 and C24 stress class timber.

STRENGTH AND SECTION SIZE CALCULATIONS

The load-bearing capacity of a timber member is a function not only of its cross-section, but also of its strength class. Readily available timbers are classified from strength class C16 and C24; these include a range of European-, UK-, Canadian- and USA-produced timbers. Whilst section size savings can be achieved using the higher-grade timbers, there is a cost premium to pay, and on small-scale projects the economy of timber section is not great (more detail will be given on this later). The C16 timber, being of less strength than the C24, results in a larger section, and in some cases the greater width of the timber can be of benefit to the non-professional, giving a greater width of timber in some instances to which to fix both battens and plasterboard ceilings and so on, whilst the tables shown below are based on C24 strength class timber and design method BS5268 (a comparison has been given in C16). For further comparison, sizes based on EC5 in both C16 and C24 have also been shown in Figure 3.6.

HOW DO WE CALCULATE THE LOADING ON THE ROOF?

The dead load – that is, that of the roof covering itself – can be obtained from the roof covering manufacturer as the 'as laid weight'; an indication of the weights of various tiles and slates can be seen in Figure 9.10. Whilst the tile loading will be expressed in kg/m^2 the dead load forces for design purposes are expressed in kN/m^2. The conversion is approximately 1/100; that is a tile weighing 75 kg/m^2 exerts a force of 0.75 kN/m^2.

The imposed load is that for snow lying on the roof; this varies, as has been stated, with geography, altitude and indeed with the roof pitch, but the latter factor is taken into account within the calculations for preparation of the tables. Snow loading is mainly related to altitude, that is, the location of the building above sea level, but also to exposure. Thus, south western counties which generally are at a lower altitude would have a snow loading of 0.75 kN/m^2, whereas more eastern and higher-altitude (not exceeding 200 m) locations vary between 0.75 kN/m^2 and 1.0 kN/m^2. The extreme north-east of England can be up to 1.07 kN/m^2 and the part of Scotland below 200 m, up to 1.25 kN/m^2. For more exact data, see snow loading maps are provided in the Trada publications described above.

TIMBER MEMBER SIZING DESIGN: AN EXAMPLE

The roof to be designed is located in the highlands of Scotland, but below 100 m altitude. The snow loading is 1.02 kN/m^2.

The ceiling joists are to carry normal loft storage only, not flooring loads. The ceiling would be considered as 12 mm plasterboard, with a load of 0.25 kN/m^2.

Ceiling Joists (Ref. Fig. 3.1)
The spacing of the ceiling joists is 400 mm, the span is 2.33 m.

Ceiling Tie Binders (Ref. Fig. 3.1)
The longest span is 2.7 m between supports, and the spacing is 2.33 m.

Purlins (Ref. Fig. 3.1)
The maximum span of the purlin between supports is 2.7 m, and the spacing of the purlins is also 2.4 m.

Rafters (Ref. Fig. 3.1)
The rafter spacing is 400 mm and maximum span is 2.4 m.

Calculating the size of timber members

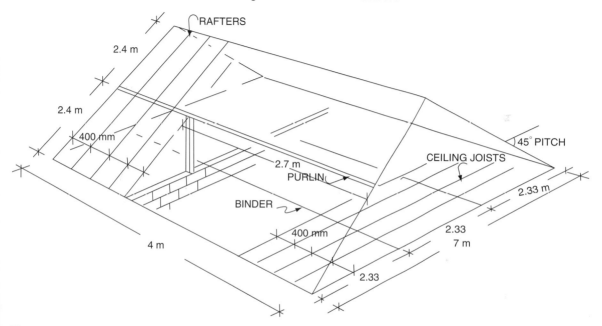

Figure 3.1 Dimensions of a roof in a worked example.

We now have all the information to find the size of timber cross-section to carry the loads imposed upon it. Refer now to the tables in Figures 3.2–3.5 below.

From Figure 3.2, for Ceiling Joists – the nearest above our required span is 2.61 m, which gives a 47 × 120 mm strength class C24 timber section. (This would be a basic 125 mm depth, but

Size of ceiling joist	Maximum clear span (m)
Breadth × Depth (mm)	
47 × 97	2.00
47 × 120	2.61
47 × 145	3.29
47 × 170	3.69
47 × 195	4.64
47 × 220	5.32

Minimum ceiling joist bearing 35 mm.
Imposed load: 0.25 kN/m^2 (concentrated load 0.9 kN).
Dead load: 0.50 kN/m^2 excluding self-weight of joist.
The above values have been compiled for guidance table by Geomex Ltd Structural Engineers: www.geomex.co.uk.
Span tables for C16- and C24-strength class solid timber members in floors, ceilings and roofs for dwellings are available from TRADA Technology at: www.trada.co.uk/bookshop.

Figure 3.2 Span tables for ceiling joists. The above table is reprinted from Guild to Building Control by Anthony Gwynne with the permission of Wiley Blackwell.

Size of binder (mm)	Spacing of binders (mm)					
	1200	1500	1800	2100	2400	2700
Breadth × Depth	Maximum clear span or hanger spacing (m)					
47 × 175	2.88	2.69	2.54	2.42[1]	2.32[1]	2.23[1]
47 × 200	3.33	3.11	2.93[1]	2.29[1]	2.67[1]	2.56[1]
75 × 175	3.43	3.21	3.04	2.90	2.78	2.67
75 × 200	3.95	3.70	3.50	3.33	3.19	3.07
75 × 225	4.47	4.18	3.95	3.76	3.60[1]	3.47[1]

Minimum ceiling binder bearing 60 mm.
Key: [1] 120 mm minimum bearing required.
Imposed load: 0.25 kN/m^2 (concentrated load 0.9 kN).
Dead load: 0.50 kN/m^2 excluding self-weight of binder.
The above values have been independently compiled for guidance table by Geomex Ltd Structural Engineers: www.geomex.co.uk.
Span tables for C16- and C24-strength class solid timber members in floors, ceilings and roofs for dwellings are available from TRADA Technology at: www.trada.co.uk/bookshop.

Figure 3.3 Span tables for ceiling binders. The above table is reprinted from Guild to Building Control by Anthony Gwynne with the permission of Wiley Blackwell.

	Slope of roof (degrees)											
	15–22.5°				22.5–30°				30–45°			
	Spacing of Purlins (mm)											
Size of purlin (mm)	1500	1800	2100	2400	1500	1800	2100	2400	1500	1800	2100	2400
B × D	Maximum clear spans (m)											
75 × 125	2.01	1.88	1.77	1.65	2.06	1.92	1.82	1.73	2.12	1.99	1.88	1.79
75 × 150	2.41	2.25	2.13	1.98	2.46	2.31	2.18	2.07	2.54	2.38	2.25	2.15
75 × 175	2.81	2.63	2.48	2.31	2.87	2.69	2.54	2.42	2.97	2.78	2.63	2.50
75 × 200	3.20	3.00	2.83	2.63	3.28	3.07	2.90	2.76	3.39	3.17	3.00	2.86
75 × 225	3.60	3.37	3.19	2.96	3.68	3.45	3.26	3.10	3.81	3.57	3.35	–

Minimum purlin bearing 80 mm.
Imposed load: 1.02 kN/m^2 (high snow load – altitudes not exceeding 100 m).
Dead load: not more than 0.75 kN/m^2 (concentrated load 0.9 kN) excluding self-weight of purlin.
The above values have been independently compiled for guidance table by Geomex Ltd Structural Engineers: www.geomex.co.uk.
Span tables for C16- and C24-strength class solid timber members in floors, ceilings and roofs for dwellings are available from TRADA Technology at: www.trada.co.uk/bookshop.

Figure 3.4 Span tables for purlins. The above table is reprinted from Guild to Building Control by Anthony Gwynne with the permission of Wiley Blackwell.

Size of rafter	Slope of roof (degrees)		
	15–22.5°	**22.5–30°**	**30–45°**
Breadth × Depth (mm)	Maximum clear span (m)		
47 × 100	2.08	2.12	2.18
47 × 125	2.74	2.79	2.87
47 × 150	3.40	3.47	3.56
47 × 195	4.59	4.68	4.81

Minimum rafter bearing 35 mm.

Imposed load: 1.02 kN/m^2 (high snow load – altitudes not exceeding 100 m).

Dead load: not more than 0.75 kN/m^2 (concentrated load 0.9 kN) excluding self-weight of rafter.

The above values have been independently compiled for guidance table by Geomex Ltd Structural Engineers: www.geomex.co.uk.

Span tables for C16- and C24-strength class solid timber members in floors, ceilings and roofs for dwellings are available from TRADA Technology at: www.trada.co.uk/bookshop.

Figure 3.5　Span tables for common rafters. The above table is reprinted from Guild to Building Control by Anthony Gwynne with the permission of Wiley Blackwell.

Roof member	BS 5268 C16	BS 5268 C24	EC5 C16	EC5 C24
Ceiling Joist size	47 × 145	47 × 120	47 × 145	47 × 120
Ceiling Binder size	75 × 200	75 × 175	72 × 175	72 × 150
Purlin size	*	75 × 200	*	75 × 200
Rafter size	47 × 125	47 × 125	47 × 125	47 × 125

*Span too large for readily available timber in this section.

Figure 3.6   Timber member size comparison between design standards and between timber strength classes.

machined to give a consistent 120 mm, thus providing a flat ceiling – rough-sawn timber can vary in depth, giving poor ceiling lines).

From Figure 3.3 Ceiling Binder. The maximum span is 2.7 m and spacing is 2.33 m. This gives a safe timber size of 75 × 175 mm in strength class C24.

From Figure 3.4, the maximum span of purlin is above 2.7 m, spacing of 2.4 m and pitch 45°, results in a purlin section size of 75 × 200 mm, strength class C24.

From Figure 3.5, the maximum span of the rafter is 2.4 m, with a spacing of 400 mm, at 45°; the nearest safe size is 47 × 125 mm; strength class C24.

For a comparison, using C16 timber and BS 5268 design methods, and C16 and C24 using EC5 design method (see Fig. 3.6).

The tables in Figures 3.2–3.5 only list 47 mm- and 75 mm-thick timber. The Trada Publications (referred to above) provide a full range of timber thicknesses available, including 38 mm, 44 mm,

47 mm, 63 mm and 72 mm. The publications also provide span tables for purlins supporting roof sheeting or decking.

COST CONSIDERATION

From the data in Figure 3.6 it can be seen that, in some instances, C24 timber results in a smaller section than C16. The C24 timber is of better quality, has fewer knots and other defects, but is more expensive than C16. Also, 38 mm- and 44 mm-thick timber could be used by reference to the Trada tables, but this may result in a deeper section; that is, the 47 × 145 mm ceiling joist may increase to 170 mm if it is 38 mm thick. As noted earlier in the chapter, the use of thicker timber – especially for ceiling joists – will be of benefit when nailing plasterboard ceilings in place. Before deciding finally on the timber size, check the costing with your timber merchant.

4 CALCULATING THE LENGTH AND CUTTING ANGLES OF TIMBER MEMBERS: DATA TABLES 5°–75°

For the purposes of explaining the use of the ready reckoner, reference should be made to the roof constructions illustrated in Figures 4.1 and 4.2. In practical terms, these constructions will cover most traditional roof forms and take account of the hip and valley infills used on trussed rafter construction, unless fully engineered trussed rafter hip and valleys have been designed. The cutting angles on all timbers for infill rafters on trussed rafter roofs, especially attic designs, can be calculated using the data tables which follow.

Before cutting any of the roof timbers, two vital pieces of information must be known. First, the span, which is the distance between the outer faces of the wall plate, and second the 'run' of the rafter, this being half the span (assuming that it is an equally pitched roof with the ridge in the middle of the span). Another vital piece of information is the pitch or the 'rise' of the roof.

Whenever possible the carpenter who is to construct the roof should at least supervise the fixing of the wall plates. These must be straight, level and parallel to each other. Where the roof has to be fitted to a 'T' or 'L' plan form of building, then the carpenter should check that the wall plates of the projections to either side of the main roof are at a true right-angle, unless of course designed

Goss's Roofing Ready Reckoner: From Timberwork to Tiles, Fifth Edition. C. N. Mindham.
© 2016 John Wiley & Sons, Ltd. Published 2016 by John Wiley & Sons, Ltd.

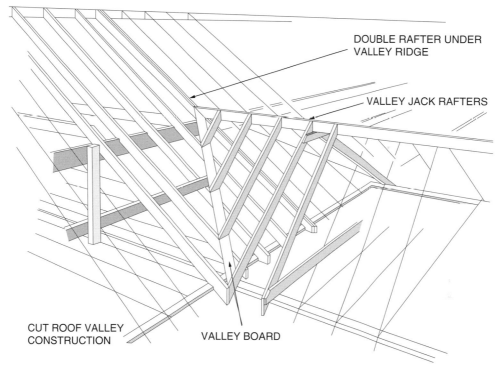

DOUBLE RAFTER UNDER
VALLEY RIDGE

VALLEY JACK RAFTERS

CUT ROOF VALLEY
CONSTRUCTION

VALLEY BOARD

Figure 4.1 Cut Roof Valley construction.

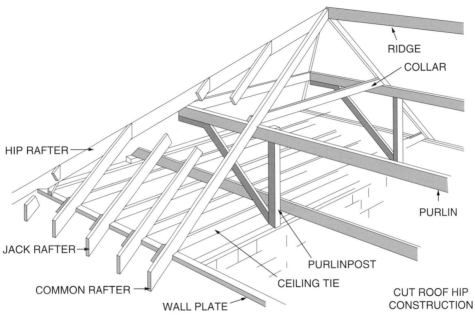

RIDGE

COLLAR

HIP RAFTER

PURLIN

JACK RAFTER

PURLINPOST

COMMON RAFTER

CEILING TIE

WALL PLATE

CUT ROOF HIP
CONSTRUCTION

Figure 4.2   Cut Roof Hip construction.

to be otherwise. Apart from checking overall dimensions with a steel tape, modern laser levels make it quick and simple to check the level of the plate very effectively, and it is this level of the wall plate which is so important to accurate roof construction.

THE PITCH

The pitch of the roof to be constructed should be clearly stated on the drawings, but if not this should be taken by protractor from the drawings, possibly extending the ceiling line and rafter line away from the point at which they meet, making it easier to get an accurate reading from the protractor. An alternative method to establish the pitch is to calculate the rise of the roof per unit of 'run'. To use the tables in this book, this must be stated in metres rise per metre run (see Fig. 4.3).

The Run of the Rafter

The run of the rafter is the horizontal distance covered by the rafter from the wall plate to the ridge (see Fig. 4.4).

The Rise of the Rafter

The rise of the rafter is the height from the top of the rafter vertically above the outside of the wall plate, to the top of the rafter at the centre line of the ridge position (see Fig. 4.4).

USING THE TABLES TO CUT A COMMON RAFTER

The use of the tables is best explained by a worked example, and to do this we will take a roof of pitch at 36° or a rise of 0.727 m per metre run, and a span of 8.46 m. Then the run:

$= 8.46 \div 2$
$= 4.23$ m

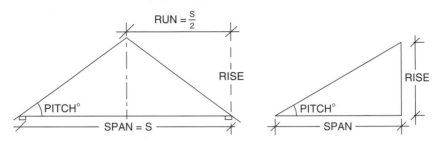

Figure 4.3 Roof Pitch Angle.

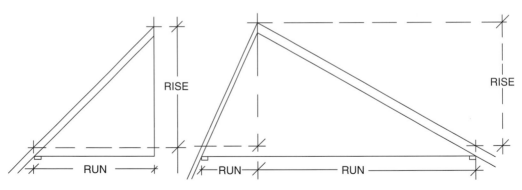

Figure 4.4 Run and Rise of Rafter.

The length of the rafter can now be calculated from the tables referring to 36° pitch. It can be seen that the length of the rafter for 1 m of run = 1.236 m; therefore, the length for 4 m of roof:

= 4 × 1.236
= 4.994 m.

The calculation for the whole rafter length now looks as follows:

4 m	=	4.944 m
0.2 m	=	0.247 m
0.03 m	=	0.0371 m
4.23 m	=	5.2281 m

Now we have the basic length of the rafter.

To calculate the *exact* length of the rafter, the length above must be reduced by half the thickness of the ridge; this can be calculated as above if perfection is required. For a ridge thickness of 40 mm, the length of run of the rafter must be reduced by 20 mm; this gives a reduction in rafter length from the tables of 0.0247 m or 24.7 mm. (Tables have to be modified by a factor of 10 because the run above is 20 mm which is 0.02 m and not 0.2 mm, as illustrated in the tables.)

We then need to add to the rafter the extra length needed to cover the overhang. This can be simply calculated in the same way by finding the length of overhang from the outside of the wall plate to the back of the fascia (see Fig. 4.5), and we will assume for the purposes of this calculation that this overhang gives a 450 mm run. Then, again by reference to the tables, it will be seen that the additional rafter length required is 556 mm. This now gives an overall rafter length as follows:

Basic rafter		5.2281 m
Add over hang	+	0.556 m
Deduct half ridge	−	0.0247 m
		5.759 m

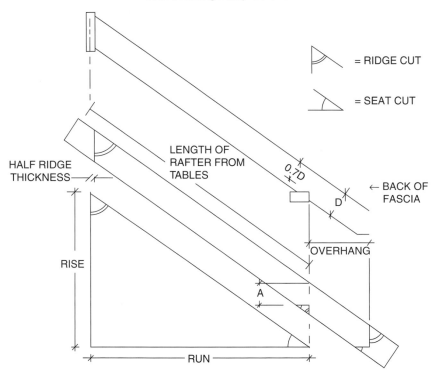

= RIDGE CUT

= SEAT CUT

HALF RIDGE
THICKNESS

LENGTH OF
RAFTER FROM
TABLES

0.7D

← BACK OF
FASCIA

D

OVERHANG

RISE

A

RUN

Figure 4.5 Rafter Length, Overhang, and Birdsmouth.

Do Not Cut Yet!

Referring now to the tables, mark the ridge bevel on one end for the ridge cut and again at the other end for the fascia cut: in this example the angle will be 54°. The distance between those marks will be the length of the rafter calculated above.

Mark the birdsmouth as shown in Figure 4.5, again using the ridge and seat bevels from the tables as indicated in the illustration.

Next mark the soffit line – this must be taken from the building design. If the soffit cut on the rafter is to be at the soffit line, the seat bevel can again be used. If the soffit line is lower than the lowest point of the rafter, then no soffit cut is required.

Now check all dimensions and angles for this first rafter, which should be regarded as the master.

Now Cut the First Rafter

Using this first rafter, check the fit to the roof and use this as a pattern to mark out all of the remaining identical rafters. Note that to give a true line for the fascia, it is common practice not to cut the fascia cut on all rafters at this stage. Leave the fascia cut on all roof members until the construction is complete; then, using a chalked line from one end of the roof to the other, mark the fascia line on the top of the rafters. From this line, using a level, a true plumb line can be marked and the rafter cut. This traditional method involves cutting the fascia cut on the roof itself, which is a time-consuming task usually done using a hand saw. Trussed rafter roofs, being prefabricated, generally have the fascia cut made at the factory. If this is the case, and allowing for some manufacturing tolerance on span, it will not be possible to line fascia cuts on both sides of the roof. There are two courses of action: (i) to re-mark and cut the foot of the trussed rafter again as outlined above; or (ii) to use packers to align the fascia onto the pre-cut feet. DO NOT be tempted to align trussed rafters to one side of the roof by aligning their rafter feet. Allowable manufacturing tolerances in a roof of the span we have been discussing could result in a variation of up to 9 mm, thus moving the ridge off the centre line, and up to a 9 mm variance between the feet of the rafters on the opposite side of the roof.

Cutting the common rafter can obviously be done by hand saw, by powered hand saw, or by using a powered compound mitre saw which can be pre-set at the ridge bevel. Then, with a saw table with the length stop at an appropriate position, all cuts will be precisely the same with no further marking required.

HIP JACK RAFTERS

The length of these members will depend on the centres at which they are to be fixed; by that we mean their spacing centre line to centre line of the thickness of the member, which should match the common rafter spacing (see Figures 2.1 and 4.6).

Continuing with the example above, the basic common rafter length was 5.2281 m, and then assuming a jack rafter spacing of 600 mm by referring to the table, it can be seen that this length must be reduced by 742 mm.

DO NOT forget to add the overhang of the common rafter; DO NOT adjust the jack rafter for the hip until a trial fit has been taken. The jack rafter meets the hip at an angle and must therefore be cut at an angle to meet the hip both horizontally and vertically, giving what is known as a compound cut. The hip, being fixed vertically in its section, gives the same bevel cut at the top of the jack rafter as was used at the ridge and this same ridge bevel can be used. However, the edge cut can be found in the tables as the 'edge bevel', and for this an example can be seen as 39°. With these two angles the top of the hip jack can be marked, and at the lower end, the common rafter master can be used to mark the fascia cut and birdsmouth.

Calculating the length and cutting angles of timber members: data tables 5°–75°

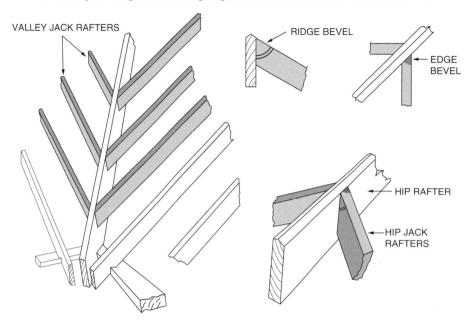

VALLEY JACK RAFTERS

RIDGE BEVEL

EDGE BEVEL

HIP RAFTER

HIP JACK RAFTERS

Figure 4.6 Hip and Valley Jack Rafters.

Cutting compound bevels by hand is a skilled task, but the powered compound mitre saw can be used to produce accurate repeatable compound cuts.

HIP RAFTERS

A full hip (i.e., that which is constructed from wall plate to ridge) will have the same rise as the common rafters, and the tables have been calculated on the assumption that the hip is on the mitre of a right-angled corner and is therefore at right-angles to the common rafters. This allows the same run as the common rafter to be used, to save further calculation.

Continuing with the example above then, the run of the hip would be 4.23 m, and by referring to the tables for the length of hip, the calculation will result in a hip length of 6.7257 m.

The seat and ridge bevels can be taken directly from the tables; in this case 27° and 63°, respectively. Care must be taken when setting out the birdsmouth to ensure that the depth of rafter (A) illustrated in Figure 4.7, equals that of the common rafter illustrated in Figure 4.5.

The mitre at the top of the hips where they meet the ridge does need the special setting of a bevel. The marking gauge is set to half the thickness of the hip and marked on the end of both faces after the plumb cut is made (see Fig. 4.7).

Backing of Hips

In a good job, the hips are backed – that is to say, a chamfer is planed both ways from the centre line on the top of the hip so that the two surfaces are in line with the planes of the roof on adjacent sides. This gives a good seating for the battens. After cutting the hip to the plumb line the same plumb bevel for marking the profile of the backing chamfers on each side of the hip may be used (see Fig. 4.7).

Calculating the length and cutting angles of timber members: data tables 5°–75°

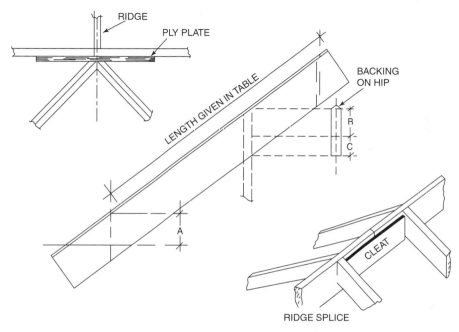

Figure 4.7 Hip and Ridge connection.

Dimension B, the length of the plumb cut of the jack rafters is measured on the top end of the hip down from the backing levels, leaving a remainder C. If C is measured along the side bevel of the purlin it gives the position of the projection under the hip (see Fig. 4.8).

VALLEY JACK RAFTERS

The tables are based on a construction which assumes a valley rafter similar to the hip rafter (see Fig. 4.6), NOT that illustrated in Figure 4.1, which is a more modern construction and one which works with a trussed rafter roof if a prefabricated valley set of frames is not provided. Returning then to the traditional cut valley, this is essentially a hip in reverse. The valley jacks decrease in length as they progress up the roof, and again would be based at similar centres to the common rafters. The same bevels as for the hip jack rafter can be used but this time on the foot of the rafter rather than the ridge as before. The common rafter ridge bevel can be used at the top.

Now, referring to Figure 4.1, the top cut on the valley jack is the same as the common rafter, but the foot of the valley rests on a flat valley board nailed on the top of the rafters of the main roof. This construction is suitable for all valleys except attic construction. The cut at the bottom of the valley jack is the same as the seat cut for the common rafter with an edge bevel equal to the pitch of the main roof. This may not be the same as the pitch of the roof on which the valley jack has to be fitted and should therefore be checked.

THE RIDGE

This roof member is usually relatively thin and can be no more than 25–38 mm. It takes little load from the roof as pairs of common rafters or valley rafters are placed directly opposite one

Calculating the length and cutting angles of timber members: data tables 5°–75°

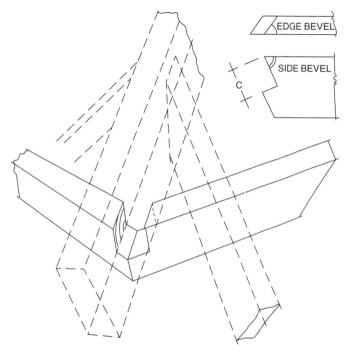

EDGE BEVEL

SIDE BEVEL

C

Figure 4.8 Abutment of Hip and Intersection of Purlins.

another across the ridge; it does act as a tie from gable to gable or hip to hip. In a roof with a gable at both ends, the length of the ridge equals the length of the wall plate and is usually built into the gables at both ends. In a hip roof with a hip at both ends, the ridge length is normally the length of the building (internally) less the width (internally). If a short packing piece is used at the end of the ridge in a hip construction to give good hip rafter supports, then this thickness must be deducted from the overall length of the ridge for each hip. The packing piece should be at least as deep as the hip rafter ridge cut, and may therefore be too deep for readily available softwood. The use of 19 mm exterior grade plywood is recommended. The ply can also be used to couple ridge pieces together in their length because even the longest lengths of timber may be too short for the overall length of the ridge itself (see Fig. 4.7). These 'cleats' should be located between rafters to avoid having to cut back the rafter by the thickness of the cleat.

PURLINS

As illustrated under the definition for purlin earlier in this book (see Fig. 2.1), the purlin can be fitted either at right-angles to the underside of the rafter or vertically. The tables given in this book show edge and side bevels for purlins set at right angles to the rafters (see Fig. 4.8). This is the traditional form of construction and may have to be followed on extension work to older buildings.

The purlin fitted vertically is preferred from a structural viewpoint because it acts as a true beam carrying the rafters. The rafters can be better fitted by birdsmouth to the purlin. The edge bevel on this type of purlin is that formed by the hip on plan view, which on a right-angled hip is 45°. The side bevel is 90°, simply a square cut.

METRIC CALCULATION TABLES

RISE OF COMMON RAFTER 0.087 m **PER METRE OF RUN**   **PITCH** 5°

BEVELS:	COMMON RAFTER	– SEAT	5
	″ ″	– RIDGE	85
	HIP OR VALLEY	– SEAT	
	″ ″ ″	– RIDGE	
	JACK RAFTER	– EDGE	
	PURLIN	– EDGE	
	″	– SIDE	

JACK RAFTERS 333 mm **CENTRES DECREASE**   (in mm to 999 and
        400 ″   ″   ″   thereafter in m)
        500 ″   ″   ″
        600 ″   ″   ″

Run of Rafter	0.1	0.2	0.3	0.4	0.5	0.6	0.7	0.8	0.9	1.0
Length of Rafter	0.1	0.201	0.301	0.401	0.502	0.602	0.703	0.803	0.903	1.004
Length of Hip										

RISE OF COMMON RAFTER 0.105 m PER METRE OF RUN   PITCH 6°

	BEVELS:	COMMON RAFTER	– SEAT	6
		" "	– RIDGE	84
		HIP OR VALLEY	– SEAT	
		" " "	– RIDGE	
		JACK RAFTER	– EDGE	
		PURLIN	– EDGE	
		"	– SIDE	

JACK RAFTERS   333 mm   **CENTRES DECREASE**   (in mm to 999 and
                   400 "   "   "   thereafter in m)
                   500 "   "   "
                   600 "   "   "

Run of Rafter	0.1	0.2	0.3	0.4	0.5	0.6	0.7	0.8	0.9	1.0
Length of Rafter	0.101	0.201	0.302	0.402	0.503	0.603	0.704	0.804	0.905	1.006
Length of Hip										

RISE OF COMMON RAFTER 0.123 m PER METRE OF RUN PITCH 7°

BEVELS: COMMON RAFTER   – SEAT 7
         "   "   – RIDGE 83
         HIP OR VALLEY   – SEAT
         "   "   "   – RIDGE
         JACK RAFTER   – EDGE
         PURLIN   – EDGE
         "   – SIDE

JACK RAFTERS 333 mm **CENTRES DECREASE** (in mm to 999 and
         400   "   "   "   thereafter in m)
         500   "   "   "
         600   "   "   "

Run of Rafter	0.1	0.2	0.3	0.4	0.5	0.6	0.7	0.8	0.9	1.0
Length of Rafter	0.101	0.202	0.302	0.403	0.504	0.605	0.705	0.806	0.907	1.008
Length of Hip										

RISE OF COMMON RAFTER   0.141 m   **PER METRE OF RUN**   **PITCH** 8°

BEVELS: COMMON RAFTER   – SEAT   8
               "   "   – RIDGE 82
         HIP OR VALLEY   – SEAT
               "   "   "   – RIDGE
         JACK RAFTER   – EDGE
         PURLIN   – EDGE
               "   – SIDE

JACK RAFTERS   333 mm **CENTRES DECREASE**   (in mm to 999 and
                   400 "   "   "   thereafter in m)
                   500 "   "   "
                   600 "   "   "

Run of Rafter	0.1	0.2	0.3	0.4	0.5	0.6	0.7	0.8	0.9	1.0
Length of Rafter	0.101	0.202	0.303	0.404	0.505	0.606	0.707	0.808	0.909	1.01
Length of Hip										

RISE OF COMMON RAFTER 0.158 m **PER METRE OF RUN**   **PITCH** 9°

BEVELS: COMMON RAFTER   – SEAT 9
                        ″   ″   – RIDGE 81
                  HIP OR VALLEY   – SEAT
                     ″   ″   ″   – RIDGE
                  JACK RAFTER   – EDGE
                  PURLIN   – EDGE
                        ″   – SIDE

JACK RAFTERS   333 mm **CENTRES DECREASE**   (in mm to 999 and
                          400 ″   ″   ″   thereafter in m)
                          500 ″   ″   ″
                          600 ″   ″   ″

	Run of Rafter	0.1	0.2	0.3	0.4	0.5	0.6	0.7	0.8	0.9	1.0
Length of Rafter		0.101	0.203	0.304	0.405	0.506	0.608	0.709	0.81	0.911	1.013
Length of Hip											

RISE OF COMMON RAFTER   0.176 m **PER METRE OF RUN**   **PITCH**   10°

BEVELS: COMMON RAFTER   – SEAT   10
                        "   "   – RIDGE   80
            HIP OR VALLEY   – SEAT
            "   "   "   – RIDGE
            JACK RAFTER   – EDGE
            PURLIN   – EDGE
                 "   – SIDE

JACK RAFTERS   333 mm   **CENTRES DECREASE**   (in mm to 999 and
                     400 "   "   "   thereafter in m)
                     500 "   "   "
                     600 "   "   "

Run of Rafter	0.1	0.2	0.3	0.4	0.5	0.6	0.7	0.8	0.9	1.0
Length of Rafter	0.102	0.203	0.305	0.406	0.508	0.609	0.711	0.812	0.914	1.015
Length of Hip										

RISE OF COMMON RAFTER   0.194 m **PER METRE OF RUN**   **PITCH**   11°

BEVELS: COMMON RAFTER   – SEAT   11
                    ″   ″   – RIDGE 79
              HIP OR VALLEY   – SEAT
              ″   ″   ″   – RIDGE
              JACK RAFTER   – EDGE
              PURLIN   – EDGE
                    ″   – SIDE

JACK RAFTERS   333 mm   **CENTRES DECREASE**   (in mm to 999 and
                     400 ″   ″   ″   thereafter in m)
                     500 ″   ″   ″
                     600 ″   ″   ″

Run of Rafter	0.1	0.2	0.3	0.4	0.5	0.6	0.7	0.8	0.9	1.0
Length of Rafter	0.102	0.204	0.306	0.407	0.509	0.611	0.713	0.815	0.917	1.019
Length of Hip										

RISE OF COMMON RAFTER 0.213 m **PER METRE OF RUN**   **PITCH**   12°

	BEVELS:	COMMON RAFTER	– SEAT	12
		" "	– RIDGE	78
		HIP OR VALLEY	– SEAT	
		" " "	– RIDGE	
		JACK RAFTER	– EDGE	
		PURLIN	– EDGE	
		"	– SIDE	

JACK RAFTERS   333 mm   **CENTRES DECREASE**   (in mm to 999 and
                  400 "   "   "   thereafter in m)
                  500 "   "   "
                  600 "   "   "

Run of Rafter	0.1	0.2	0.3	0.4	0.5	0.6	0.7	0.8	0.9	1.0
Length of Rafter	0.102	0.204	0.307	0.409	0.511	0.613	0.716	0.818	0.92	1.022
Length of Hip										

RISE OF COMMON RAFTER 0.231 m **PER METRE OF RUN**   **PITCH** 13°

BEVELS: COMMON RAFTER   – SEAT 13
                "   "   – RIDGE 77
           HIP OR VALLEY   – SEAT
           "   "   "   – RIDGE
           JACK RAFTER   – EDGE
           PURLIN   – EDGE
           "   – SIDE

JACK RAFTERS   333 mm **CENTRES DECREASE**   (in mm to 999 and
               400 "   "   "   thereafter in m)
               500 "   "   "
               600 "   "   "

Run of Rafter	0.1	0.2	0.3	0.4	0.5	0.6	0.7	0.8	0.9	1.0
Length of Rafter	0.103	0.205	0.308	0.411	0.513	0.616	0.718	0.821	0.924	1.026
Length of Hip										

RISE OF COMMON RAFTER 0.249 m **PER METRE OF RUN** **PITCH** 14°

BEVELS: COMMON RAFTER   – SEAT   14
            "   "   – RIDGE 76
          HIP OR VALLEY   – SEAT
            "   "   "   – RIDGE
          JACK RAFTER   – EDGE
          PURLIN   – EDGE
            "   – SIDE

JACK RAFTERS 333 mm **CENTRES DECREASE**   (in mm to 999 and
                  400 "   "   "   thereafter in m)
                  500 "   "   "
                  600 "   "   "

Run of Rafter	0.1	0.2	0.3	0.4	0.5	0.6	0.7	0.8	0.9	1.0
Length of Rafter	0.103	0.206	0.309	0.412	0.515	0.618	0.721	0.824	0.928	1.031
Length of Hip										

Calculating the length and cutting angles of timber members: data tables 5°–75°

RISE OF COMMON RAFTER 0.268 m **PER METRE OF RUN**   **PITCH** 15°

BEVELS: COMMON RAFTER   – SEAT   15
                   ″   ″   – RIDGE 75
          HIP OR VALLEY   – SEAT
          ″   ″   ″   – RIDGE
          JACK RAFTER   – EDGE
          PURLIN   – EDGE
                   ″   – SIDE

JACK RAFTERS   333 mm   **CENTRES DECREASE**   (in mm to 999 and
                   400 ″   ″   ″   thereafter in m)
                   500 ″   ″   ″
                   600 ″   ″   ″

Run of Rafter	0.1	0.2	0.3	0.4	0.5	0.6	0.7	0.8	0.9	1.0
Length of Rafter	0.104	0.207	0.311	0.414	0.518	0.621	0.725	0.828	0.932	1.035
Length of Hip										

RISE OF COMMON RAFTER 0.287 m **PER METRE OF RUN**

PITCH 16°
(Grecian pitch)

BEVELS: COMMON RAFTER   – SEAT 16
                    "   "   – RIDGE 74
              HIP OR VALLEY   – SEAT 11.5
                "   "   "   – RIDGE 78.5
              JACK RAFTER   – EDGE 44
              PURLIN   – EDGE 46
                    "   – SIDE   74.5

JACK RAFTERS 333 mm **CENTRES DECREASE** 346 (in mm to 999 and
                  400 "   "   "   416 thereafter in m)
                  500 "   "   "   520
                  600 "   "   "   624

Run of Rafter	0.1	0.2	0.3	0.4	0.5	0.6	0.7	0.8	0.9	1.0
Length of Rafter	0.104	0.208	0.312	0.416	0.52	0.624	0.728	0.832	0.936	1.04
Length of Hip	0.144	0.289	0.433	0.577	0.722	0.866	1.01	1.154	1.299	1.443

RISE OF COMMON RAFTER 0.306 m **PER METRE OF RUN**   **PITCH** 17°

BEVELS:	COMMON RAFTER	– SEAT	17
	" "	– RIDGE	73
	HIP OR VALLEY	– SEAT	12
	" " "	– RIDGE	78
	JACK RAFTER	– EDGE	43.5
	PURLIN	– EDGE	46.5
	"	– SIDE	73.5

JACK RAFTERS 333 mm **CENTRES DECREASE** 348 (in mm to 999 and
400 "   "   "   418 thereafter in m)
500 "   "   "   522
600 "   "   "   627

Run of Rafter	0.1	0.2	0.3	0.4	0.5	0.6	0.7	0.8	0.9	1.0
Length of Rafter	0.105	0.209	0.314	0.418	0.523	0.627	0.732	0.837	0.941	1.046
Length of Hip	0.145	0.289	0.434	0.579	0.723	0.868	1.013	1.158	1.302	1.447

RISE OF COMMON RAFTER 0.315 m **PER METRE OF RUN**   **PITCH** $17\frac{1}{2}°$

BEVELS:	COMMON RAFTER	– SEAT	17.5
	" "	– RIDG	72.5
	HIP OR VALLEY	– SEA	12.5
	" " "	– RIDGE	77.5
	JACK RAFTER	– EDGE	43.5
	PURLIN	– EDGE	46.5
	"	– SIDE	73.5

JACK RAFTERS 333 mm **CENTRES DECREASE** 349 (in mm to 999 and
                 400 "   "   "   420 thereafter in m)
                 500 "   "   "   524
                 600 "   "   "   629

Run of Rafter	0.1	0.2	0.3	0.4	0.5	0.6	0.7	0.8	0.9	1.0
Length of Rafter	0.105	0.210	0.315	0.419	0.524	0.629	0.734	0.839	0.944	1.049
Length of Hip	0.145	0.29	0.435	0.579	0.724	0.829	1.014	1.159	1.304	1.449

Calculating the length and cutting angles of timber members: data tables 5°–75°

RISE OF COMMON RAFTER 0.325 m **PER METRE OF RUN**   **PITCH**   18°

BEVELS: COMMON RAFTER   – SEAT   18
                              ″   ″   – RIDGE   72
                HIP OR VALLEY   – SEAT   13
                  ″   ″   ″   – RIDGE   77
                JACK RAFTER   – EDGE   43.5
                PURLIN   – EDGE   46.5
                     ″   – SIDE   73

JACK RAFTERS   333 mm   **CENTRES DECREASE**   350   (in mm to 999 and
                     400 ″   ″   ″   421   thereafter in m)
                     500 ″   ″   ″   526
                     600 ″   ″   ″   631

Run of Rafter	0.1	0.2	0.3	0.4	0.5	0.6	0.7	0.8	0.9	1.0
Length of Rafter	0.105	0.21	0.315	0.421	0.526	0.631	0.736	0.841	0.946	1.051
Length of Hip	0.145	0.29	0.435	0.58	0.726	0.871	1.016	1.161	1.306	1.451

RISE OF COMMON RAFTER   0.344 m **PER METRE OF RUN**   **PITCH**   19°

BEVELS:	COMMON RAFTER	– SEAT	19
	〃 　　　〃	– RIDGE	71
	HIP OR VALLEY	– SEAT	13.5
	〃 　〃 　〃	– RIDGE	76.5
	JACK RAFTER	– EDGE	43.5
	PURLIN	– EDGE	46.5
	〃	– SIDE	72

JACK RAFTERS   333 mm   **CENTRES DECREASE**   352   (in mm to 999 and
　　　　　　　400 〃　　　　　〃　　　　　〃　　　423   thereafter in m)
　　　　　　　500 〃　　　　　〃　　　　　〃　　　529
　　　　　　　600 〃　　　　　〃　　　　　〃　　　635

Run of Rafter	0.1	0.2	0.3	0.4	0.5	0.6	0.7	0.8	0.9	1.0
Length of Rafter	0.106	0.212	0.317	0.423	0.529	0.635	0.74	0.846	0.952	1.058
Length of Hip	0.146	0.291	0.437	0.582	0.728	0.873	1.019	1.164	1.31	1.456

RISE OF COMMON RAFTER 0.364 m **PER METRE OF RUN**   **PITCH** 20°

BEVELS:	COMMON RAFTER	– SEAT	20	
	" "	– RIDGE	70	
	HIP OR VALLEY	– SEAT	14.5	
	" " "	– RIDGE	75.5	
	JACK RAFTER	– EDGE	43	
	PURLIN	– EDGE	47	
	"	– SIDE	71	

JACK RAFTERS 333 mm **CENTRES DECREASE** 354 (in mm to 999 and
             400 "   "   "   426 thereafter in m)
             500 "   "   "   532
             600 "   "   "   638

Run of Rafter	0.1	0.2	0.3	0.4	0.5	0.6	0.7	0.8	0.9	1.0
Length of Rafter	0.106	0.213	0.319	0.426	0.532	0.639	0.745	0.851	0.958	1.064
Length of Hip	0.146	0.292	0.438	0.584	0.73	0.876	1.022	1.168	1.314	1.46

RISE OF COMMON RAFTER 0.384 m **PER METRE OF RUN** **PITCH** 21°

BEVELS: COMMON RAFTER – SEAT 21
"   "   – RIDGE 69
HIP OR VALLEY – SEAT 15
"   "   "   – RIDGE 75
JACK RAFTER – EDGE 43
PURLIN – EDGE 47
"   – SIDE 70.5

JACK RAFTERS 333 mm **CENTRES DECREASE** 357 (in mm to 999 and
400 "   "   "   428 thereafter in m)
500 "   "   "   536
600 "   "   "   643

Run of Rafter	0.1	0.2	0.3	0.4	0.5	0.6	0.7	0.8	0.9	1.0
Length of Rafter	0.107	0.214	0.321	0.428	0.536	0.643	0.75	0.857	0.964	1.071
Length of Hip	0.147	0.293	0.44	0.586	0.733	0.879	1.026	1.172	1.319	1.465

RISE OF COMMON RAFTER 0.404 m **PER METRE OF RUN**   **PITCH** 22°

BEVELS: COMMON RAFTER   – SEAT   22
                    "   "   – RIDGE 68
                 HIP OR VALLEY   – SEAT   16
                 "   "   "   – RIDGE 74
                 JACK RAFTER   – EDGE 43
                 PURLIN   – EDGE 47
                     "   – SIDE 69.5

JACK RAFTERS   333 mm **CENTRES DECREASE**   359   (in mm to 999 and
                   400 "   "   "   432   thereafter in m)
                   500 "   "   "   540
                   600 "   "   "   647

Run of Rafter	0.1	0.2	0.3	0.4	0.5	0.6	0.7	0.8	0.9	1.0
Length of Rafter	0.108	0.216	0.324	0.431	0.539	0.647	0.755	0.863	0.971	1.079
Length of Hip	0.147	0.294	0.441	0.588	0.736	0.883	1.03	1.177	1.324	1.471

RISE OF COMMON RAFTER 0.414 m **PER METRE OF RUN**   **PITCH** $22\frac{1}{2}°$

BEVELS: COMMON RAFTER   – SEAT   22.5
                  "   "   – RIDGE 67.5
           HIP OR VALLEY   – SEAT   16.25
             "   "   "   – RIDGE 73.75
           JACK RAFTER   – EDGE   42.75
           PURLIN   – EDGE   47.5
               "   – SIDE   69.0

JACK RAFTERS   333 mm   **CENTRES DECREASE**   361   (in mm to 999 and
                   400 "   "   "   433   thereafter in m)
                   500 "   "   "   542
                   600 "   "   "   650

Run of Rafter	0.1	0.2	0.3	0.4	0.5	0.6	0.7	0.8	0.9	1.0
Length of Rafter	0.108	0.216	0.325	0.433	0.541	0.649	0.758	0.866	0.974	1.082
Length of Hip	0.147	0.294	0.442	0.589	0.737	0.884	1.032	1.179	1.326	1.473

RISE OF COMMON RAFTER 0.424 m **PER METRE OF RUN**   **PITCH** 23°

BEVELS:	COMMON RAFTER	– SEAT	23
	" "	– RIDGE	67
	HIP OR VALLEY	– SEAT	16.5
	" " "	– RIDGE	73.5
	JACK RAFTER	– EDGE	42.5
	PURLIN	– EDGE	47.5
	"	– SIDE	68.5

JACK RAFTERS 333 mm **CENTRES DECREASE**   362 (in mm to 999 and
           400 "   "   "   434 thereafter in m)
           500 "   "   "   543
           600 "   "   "   652

Run of Rafter	0.1	0.2	0.3	0.4	0.5	0.6	0.7	0.8	0.9	1.0
Length of Rafter	0.109	0.217	0.326	0.435	0.543	0.652	0.76	0.869	0.978	1.086
Length of Hip	0.148	0.295	0.443	0.591	0.739	0.886	1.034	1.181	1.329	1.477

RISE OF COMMON RAFTER 0.445 m **PER METRE OF RUN**   **PITCH** 24°
(Roman pitch)

BEVELS: COMMON RAFTER   – SEAT   24
             ″   ″   – RIDGE 66
          HIP OR VALLEY   – SEAT   17.5
           ″   ″   ″   – RIDGE 72.5
          JACK RAFTER   – EDGE   42.5
          PURLIN   – EDGE   47.5
             ″   – SIDE   68

JACK RAFTERS   333 mm **CENTRES DECREASE**   365   (in mm to 999 and
                   400 ″   ″   ″   438   thereafter in m)
                   500 ″   ″   ″   548
                   600 ″   ″   ″   657

Run of Rafter	0.1	0.2	0.3	0.4	0.5	0.6	0.7	0.8	0.9	1.0
Length of Rafter	0.109	0.219	0.328	0.438	0.547	0.657	0.766	0.876	0.985	1.095
Length of Hip	0.148	0.297	0.445	0.593	0.741	0.89	1.038	1.186	1.334	1.483

Calculating the length and cutting angles of timber members: data tables 5°–75°

RISE OF COMMON RAFTER 0.466 m PER METRE OF RUN   PITCH 25°

BEVELS: COMMON RAFTER   – SEAT 25
                     "   "   – RIDGE 65
              HIP OR VALLEY   – SEAT 18
              " " "   – RIDGE 72
              JACK RAFTER   – EDGE 42
              PURLIN   – EDGE 48
              "   – SIDE 67

JACK RAFTERS 333 mm **CENTRES DECREASE** 367 (in mm to 999 and
400 "   "   "   441 thereafter in m)
500 "   "   "   552
600 "   "   "   662

Run of Rafter	0.1	0.2	0.3	0.4	0.5	0.6	0.7	0.8	0.9	1.0
Length of Rafter	0.11	0.221	0.331	0.441	0.552	0.662	0.772	0.883	0.993	1.103
Length of Hip	0.149	0.298	0.447	0.596	0.745	0.893	1.042	1.191	1.34	1.489

RISE OF COMMON RAFTER 0.5 m **PER METRE OF RUN**   **PITCH** 26° 34′
(Ouarter pitch)

BEVELS: COMMON RAFTER – SEAT 26.5
         ″   ″   – RIDGE 63.5
         HIP OR VALLEY   – SEAT 19.5
         ″   ″   ″   – RIDGE 70.5
         JACK RAFTER   – EDGE 42
         PURLIN   – EDGE 48
         ″   – SIDE 66

JACK RAFTERS 333 mm **CENTRES DECREASE** 372 (in mm to 999 and
                  400 ″   ″   ″   447 thereafter in m)
                  500 ″   ″   ″   559
                  600 ″   ″   ″   671

Run of Rafter	0.1	0.2	0.3	0.4	0.5	0.6	0.7	0.8	0.9	1.0
Length of Rafter	0.112	0.224	0.335	0.447	0.559	0.671	0.783	0.894	1.006	1.118
Length of Hip	0.15	0.3	0.45	0.6	0.75	0.9	1.05	1.2	1.35	1.5

RISE OF COMMON RAFTER 0.521 m PER METRE OF RUN   PITCH $22\frac{1}{2}°$

BEVELS:	COMMON RAFTER	– SEAT	27.5
	" "	– RIDGE	62.5
	HIP OR VALLEY	– SEAT	20
	" " "	– RIDGE	70
	JACK RAFTER	– EDGE	41.75
	PURLIN	– EDGE	48.5
	"	– SIDE	65

JACK RAFTERS 333 mm **CENTRES DECREASE** 375 (in mm to 999 and
                400 "   "   "   451 thereafter in m)
                500 "   "   "   563
                600 "   "   "   677

Run of Rafter	0.1	0.2	0.3	0.4	0.5	0.6	0.7	0.8	0.9	1.0
Length of Rafter	0.113	0.225	0.338	0.451	0.564	0.676	0.789	0.902	1.015	1.127
Length of Hip	0.151	0.301	0.452	0.603	0.754	0.904	1.054	1.206	1.365	1.507

RISE OF COMMON RAFTER 0.532 m **PER METRE OF RUN**   **PITCH** 28°

BEVELS: COMMON RAFTER   – SEAT   28
                "   "   – RIDGE 62
          HIP OR VALLEY   – SEAT   20.5
           "   "   "   – RIDGE 69.5
          JACK RAFTER   – EDGE   41.5
          PURLIN   – EDGE   48.5
           "   – SIDE   65

JACK RAFTERS 333 mm **CENTRES DECREASE** 377   (in mm to 999 and
                  400 "   "   "   453   thereafter in m)
                  500 "   "   "   566
                  600 "   "   "   680

Run of Rafter	0.1	0.2	0.3	0.4	0.5	0.6	0.7	0.8	0.9	1.0
Length of Rafter	0.113	0.227	0.34	0.453	0.566	0.68	0.793	0.906	1.019	1.133
Length of Hip	0.151	0.302	0.453	0.603	0.754	0.905	1.056	1.207	1.358	1.511

RISE OF COMMON RAFTER 0.544 m **PER METRE OF RUN**   **PITCH** 29°

BEVELS:	COMMON RAFTER	– SEAT	29
	" "	– RIDGE	61
	HIP OR VALLEY	– SEAT	21.5
	" " "	– RIDGE	68.5
	JACK RAFTER	– EDGE	41
	PURLIN	– EDGE	49
	"	– SIDE	64

JACK RAFTERS   333 mm   **CENTRES DECREASE**   381   (in mm to 999 and
                      400 "   "   "   457   thereafter in m)
                      500 "   "   "   572
                      600 "   "   "   686

Run of Rafter	0.1	0.2	0.3	0.4	0.5	0.6	0.7	0.8	0.9	1.0
Length of Rafter	0.114	0.229	0.343	0.457	0.572	0.686	0.8	0.914	1.029	1.143
Length of Hip	0.152	0.304	0.456	0.608	0.759	0.912	1.063	1.215	1.367	1.519

RISE OF COMMON RAFTER 0.577 m **PER METRE OF RUN**   **PITCH** 30°

BEVELS: COMMON RAFTER   – SEAT   30
          "   "   – RIDGE 60
          HIP OR VALLEY   – SEAT   22
          "   "   "   – RIDGE 68
          JACK RAFTER   – EDGE   41
          PURLIN   – EDGE   49
          "   – SIDE   63.5

JACK RAFTERS 333 mm **CENTRES DECREASE**   385   (in mm to 999 and
                  400 "   "   "   462   thereafter in m)
                  500 "   "   "   577
                  600 "   "   "   693

Run of Rafter	0.1	0.2	0.3	0.4	0.5	0.6	0.7	0.8	0.9	1.0
Length of Rafter	0.116	0.231	0.346	0.462	0.577	0.693	0.808	0.924	1.039	1.155
Length of Hip	0.153	0.306	0.458	0.611	0.764	0.917	1.069	1.222	1.375	1.528

RISE OF COMMON RAFTER 0.601 m PER METRE OF RUN   PITCH 31°

	BEVELS:	COMMON RAFTER	– SEAT	31
		" "	– RIDGE	59
		HIP OR VALLEY	– SEAT	23
		" " "	– RIDGE	67
		JACK RAFTER	– EDGE	40.5
		PURLIN	– EDGE	49.5
		"	– SIDE	62.5

JACK RAFTERS 333 mm **CENTRES DECREASE** 389 (in mm to 999 and

400	"	"	"	467	thereafter in m)
500	"	"	"	584	
600	"	"	"	700	

Run of Rafter	0.1	0.2	0.3	0.4	0.5	0.6	0.7	0.8	0.9	1.0
Length of Rafter	0.117	0.233	0.35	0.467	0.583	0.7	0.817	0.933	1.05	1.167
Length of Hip	0.154	0.307	0.461	0.615	0.768	0.922	1.076	1.229	1.383	1.537

RISE OF COMMON RAFTER 0.625 m **PER METRE OF RUN**   **PITCH** 32°

BEVELS: COMMON RAFTER   – SEAT   32
                 ″   ″   – RIDGE 58
             HIP OR VALLEY   – SEAT   24
                 ″   ″   ″   – RIDGE 66
             JACK RAFTER   – EDGE   40.5
             PURLIN   – EDGE   49.5
                 ″   – SIDE   62

JACK RAFTERS 333 mm **CENTRES DECREASE**   393   (in mm to 999 and
                  400 ″   ″   ″   472   thereafter in m)
                  500 ″   ″   ″   590
                  600 ″   ″   ″   707

Run of Rafter	0.1	0.2	0.3	0.4	0.5	0.6	0.7	0.8	0.9	1.0
Length of Rafter	0.118	0.239	0.354	0.472	0.59	0.708	0.825	0.943	1.061	1.179
Length of Hip	0.155	0.309	0.464	0.618	0.773	0.928	1.082	1.237	1.391	1.546

RISE OF COMMON RAFTER 0.637 m **PER METRE OF RUN**   **PITCH** $32\frac{1}{2}°$

BEVELS:	COMMON RAFTER	– SEAT	32.5
	" "	– RIDGE	57.5
	HIP OR VALLEY	– SEAT	24.25
	" " "	– RIDGE	65.75
	JACK RAFTER	– EDGE	40.25
	PURLIN	– EDGE	49.75
	"	– SIDE	61.5

JACK RAFTERS 333 mm **CENTRES DECREASE** 395 (in mm to 999 and
             400 "   "   "   475 thereafter in m)
             500 "   "   "   593
             600 "   "   "   711

Run of Rafter	0.1	0.2	0.3	0.4	0.5	0.6	0.7	0.8	0.9	1.0
Length of Rafter	0.119	0.237	0.356	0.474	0.593	0.711	0.830	0.949	1.067	1.186
Length of Hip	0.155	0.310	0.466	0.620	0.776	0.930	1.086	1.241	1.396	1.551

RISE OF COMMON RAFTER 0.649 m PER METRE OF RUN   PITCH 33°

BEVELS: COMMON RAFTER   – SEAT   33
         "   "   – RIDGE 57
         HIP OR VALLEY   – SEAT   24.5
         "   "   "   – RIDGE 65.5
         JACK RAFTER   – EDGE   40
         PURLIN   – EDGE   50
         "   – SIDE   61.5

JACK RAFTERS 333 mm **CENTRES DECREASE** 397 (in mm to 999 and
                 400 "   "   "   477 thereafter in m)
                 500 "   "   "   596
                 600 "   "   "   715

Run of Rafter	0.1	0.2	0.3	0.4	0.5	0.6	0.7	0.8	0.9	1.0
Length of Rafter	0.119	0.238	0.358	0.48	0.596	0.715	0.835	0.954	1.073	1.192
Length of Hip	0.156	0.311	0.467	0.623	0.778	0.934	1.089	1.245	1.401	1.556

Calculating the length and cutting angles of timber members: data tables 5°–75°

RISE OF COMMON RAFTER 0.666 m PER METRE OF RUN

PITCH 33° 40′
(Third pitch)

BEVELS:	COMMON RAFTER	– SEAT	33.5
	" "	– RIDGE	56.5
	HIP OR VALLEY	– SEAT	25
	" " "	– RIDGE	65
	JACK RAFTER	– EDGE	40
	PURLIN	– EDGE	50
	"	– SIDE	61

JACK RAFTERS 333 mm **CENTRES DECREASE** 397 (in mm to 999 and
400 " " " 481 thereafter in m)
500 " " " 601
600 " " " 721

Run of Rafter	0.1	0.2	0.3	0.4	0.5	0.6	0.7	0.8	0.9	1.0
Length of Rafter	0.12	0.24	0.361	0.481	0.601	0.721	0.841	0.961	1.082	1.202
Length of Hip	0.157	0.313	0.47	0.626	0.782	0.938	1.094	1.251	1.408	1.563

67

RISE OF COMMON RAFTER 0.7 m **PER METRE OF RUN**   **PITCH** 35°

BEVELS: COMMON RAFTER   – SEAT   35
                          "   "   – RIDGE 55
                HIP OR VALLEY   – SEAT   26.5
                  "   "   "   – RIDGE 63.5
                JACK RAFTER   – EDGE   39.5
                PURLIN   – EDGE   50.5
                    "   – SIDE   60

JACK RAFTERS   333 mm   **CENTRES DECREASE**   407   (in mm to 999 and
                        400 "   "   "   488   thereafter in m)
                        500 "   "   "   611
                        600 "   "   "   733

Run of Rafter	0.1	0.2	0.3	0.4	0.5	0.6	0.7	0.8	0.9	1.0
Length of Rafter	0.122	0.244	0.366	0.488	0.61	0.732	0.855	0.977	1.099	1.221
Length of Hip	0.158	0.316	0.473	0.631	0.789	0.947	1.105	1.262	1.42	1.578

RISE OF COMMON RAFTER 0.727 m **PER METRE OF RUN**   **PITCH** 36°

BEVELS: COMMON RAFTER   – SEAT   36
                 ″   ″   – RIDGE 54
           HIP OR VALLEY   – SEAT   27
             ″   ″   ″   – RIDGE 63
           JACK RAFTER   – EDGE   39
           PURLIN   – EDGE   51
             ″   – SIDE   59.5

JACK RAFTERS 333 mm **CENTRES DECREASE** 412 (in mm to 999 and
                  400 ″   ″   ″   494   thereafter in m)
                  500 ″   ″   ″   618
                  600 ″   ″   ″   742

Run of Rafter	0.1	0.2	0.3	0.4	0.5	0.6	0.7	0.8	0.9	1.0
Length of Rafter	0.124	0.247	0.371	0.494	0.618	0.742	0.865	0.989	1.112	1.236
Length of Hip	0.159	0.318	0.477	0.636	0.795	0.954	1.113	1.272	1.431	1.59

RISE OF COMMON RAFTER 0.754 m **PER METRE OF RUN**　　　　　**PITCH**　37°

BEVELS:　COMMON RAFTER　　– SEAT　37
　　　　　　　"　　　　　"　　　– RIDGE 53
　　　　　HIP OR VALLEY　　　– SEAT　28
　　　　　　"　　"　　"　　　– RIDGE 62
　　　　　JACK RAFTER　　　　– EDGE　38.5
　　　　　PURLIN　　　　　　　– EDGE　51.5
　　　　　　"　　　　　　　　– SIDE　59

JACK RAFTERS　333 mm　**CENTRES DECREASE**　417　(in mm to 999 and
　　　　　　　　　400　"　　　"　　　　　"　　501　thereafter in m)
　　　　　　　　　500　"　　　"　　　　　"　　626
　　　　　　　　　600　"　　　"　　　　　"　　751

	0.1	0.2	0.3	0.4	0.5	0.6	0.7	0.8	0.9	1.0
Run of Rafter	0.1	0.2	0.3	0.4	0.5	0.6	0.7	0.8	0.9	1.0
Length of Rafter	0.125	0.25	0.376	0.501	0.626	0.751	0.876	1.002	1.127	1.252
Length of Hip	0.16	0.32	0.481	0.641	0.801	0.961	1.122	1.282	1.442	1.602

RISE OF COMMON RAFTER 0.767 m **PER METRE OF RUN**   **PITCH** $37\frac{1}{2}°$

BEVELS: COMMON RAFTER   – SEAT   37.5
            "   "   – RIDGE 52.5
       HIP OR VALLEY   – SEAT   28.5
       "   "   "   – RIDGE 61.5
       JACK RAFTER   – EDGE 38.25
       PURLIN   – EDGE 51.75
       "   – SIDE   58.5

JACK RAFTERS 333 mm **CENTRES DECREASE** 420 (in mm to 999 and
     400 "   "   "   505   thereafter in m)
     500 "   "   "   630
     600 "   "   "   757

Run of Rafter	0.1	0.2	0.3	0.4	0.5	0.6	0.7	0.8	0.9	1.0
Length of Rafter	0.126	0.252	0.378	0.504	0.630	0.756	0.882	1.008	1.134	1.260
Length of Hip	0.161	0.322	0.482	0.636	0.804	1.965	1.126	1.286	1.448	1.609

RISE OF COMMON RAFTER 0.781 m **PER METRE OF RUN**   **PITCH** 38°

BEVELS:	COMMON RAFTER	– SEAT	38	
	" "	– RIDGE	52	
	HIP OR VALLEY	– SEAT	29	
	" " "	– RIDGE	61	
	JACK RAFTER	– EDGE	38	
	PURLIN	– EDGE	52	
	"	– SIDE	58.5	

JACK RAFTERS	333 mm	**CENTRES DECREASE**		423	(in mm to 999 and
	400 "	"	"	508	thereafter in m)
	500 "	"	"	635	
	600 "	"	"	761	

Run of Rafter	0.1	0.2	0.3	0.4	0.5	0.6	0.7	0.8	0.9	1.0
Length of Rafter	0.127	0.254	0.381	0.508	0.635	0.761	0.888	1.015	1.142	1.269
Length of Hip	0.162	0.323	0.485	0.646	0.808	0.969	1.131	1.293	1.454	1.616

RISE OF COMMON RAFTER 0.81 m PER METRE OF RUN   PITCH   39°

BEVELS:	COMMON RAFTER	– SEAT	39
	" "	– RIDGE	51
	HIP OR VALLEY	– SEAT	30
	" " "	– RIDGE	60
	JACK RAFTER	– EDGE	38
	PURLIN	– EDGE	52
	"	– SIDE	58

JACK RAFTERS 333 mm **CENTRES DECREASE** 429 (in mm to 999 and
                 400 "   "   "   515   thereafter in m)
                 500 "   "   "   644
                 600 "   "   "   772

Run of Rafter	0.1	0.2	0.3	0.4	0.5	0.6	0.7	0.8	0.9	1.0
Length of Rafter	0.129	0.257	0.386	0.515	0.643	0.772	0.901	1.029	1.158	1.287
Length of Hip	0.163	0.326	0.489	0.652	0.815	0.978	1.141	1.304	1.467	1.63

RISE OF COMMON RAFTER 0.839 m **PER METRE OF RUN**   **PITCH** 40°

BEVELS: COMMON RAFTER – SEAT 40
         "   "   – RIDGE 50
       HIP OR VALLEY – SEAT 30.5
       " "   "   – RIDGE 59.5
       JACK RAFTER – EDGE 37.5
       PURLIN – EDGE 52.5
       "   – SIDE 57.5

JACK RAFTERS 333 mm **CENTRES DECREASE** 435 (in mm to 999 and
       400 "   "   "   522 thereafter in m)
       500 "   "   "   653
       600 "   "   "   783

Run of Rafter	0.1	0.2	0.3	0.4	0.5	0.6	0.7	0.8	0.9	1.0
Length of Rafter	0.131	0.261	0.392	0.522	0.653	0.783	0.914	1.044	1.175	1.305
Length of Hip	0.164	0.329	0.493	0.658	0.822	0.987	1.151	1.316	1.48	1.644

Calculating the length and cutting angles of timber members: data tables 5°–75°

RISE OF COMMON RAFTER 0.869 m PER METRE OF RUN   PITCH 41°

BEVELS: COMMON RAFTER   – SEAT   41
               "   "   – RIDGE 49
            HIP OR VALLEY   – SEAT   31.5
            "   "   "   – RIDGE 58.5
            JACK RAFTER   – EDGE   37
            PURLIN   – EDGE   53
            "   – SIDE   56.5

JACK RAFTERS 333 mm **CENTRES DECREASE** 441 (in mm to 999 and
          400 "   "   "   530 thereafter in m)
          500 "   "   "   663
          600 "   "   "   795

Run of Rafter	0.1	0.2	0.3	0.4	0.5	0.6	0.7	0.8	0.9	1.0
Length of Rafter	0.133	0.265	0.398	0.53	0.663	0.795	0.928	1.06	1.193	1.325
Length of Hip	0.166	0.332	0.498	0.664	0.83	0.996	1.162	1.328	1.494	1.66

RISE OF COMMON RAFTER 0.9 m **PER METRE OF RUN**   **PITCH** 42°

BEVELS:	COMMON RAFTER	– SEAT	42
	" "	– RIDGE	48
	HIP OR VALLEY	– SEAT	32.5
	" " "	– RIDGE	57.5
	JACK RAFTER	– EDGE	36.5
	PURLIN	– EDGE	53.5
	"	– SIDE	56

JACK RAFTERS 333 mm **CENTRES DECREASE** 448 (in mm to 999 and
                400 "   "   " 538 thereafter in m)
                500 "   "   " 673
                600 "   "   " 808

Run of Rafter	0.1	0.2	0.3	0.4	0.5	0.6	0.7	0.8	0.9	1.0
Length of Rafter	0.135	0.269	0.404	0.538	0.673	0.807	0.942	1.097	1.211	1.346
Length of Hip	0.168	0.335	0.503	0.671	0.838	1.006	1.173	1.341	1.509	1.677

RISE OF COMMON RAFTER 0.916 m **PER METRE OF RUN**   **PITCH** $42\frac{1}{2}°$

BEVELS: COMMON RAFTER   – SEAT   42.5
                           "   "   – RIDGE 47.5
                  HIP OR VALLEY   – SEAT   33
                  "   "   "   – RIDGE 57
                  JACK RAFTER   – EDGE   36.25
                  PURLIN   – EDGE   53.75
                        "   – SIDE   55.75

JACK RAFTERS 333 mm **CENTRES DECREASE**   452   (in mm to 999 and
                    400 "   "   "   543   thereafter in m)
                    500 "   "   "   679
                    600 "   "   "   815

Run of Rafter	0.1	0.2	0.3	0.4	0.5	0.6	0.7	0.8	0.9	1.0
Length of Rafter	0.137	0.271	0.406	0.543	0.678	0.814	0.949	1.085	1.221	1.356
Length of Hip	0.170	0.337	0.505	0.674	0.842	1.011	1.179	1.348	1.569	1.685

RISE OF COMMON RAFTER 0.933 m **PER METRE OF RUN**   **PITCH** 43°

BEVELS: COMMON RAFTER   – SEAT   43
                 "   "   – RIDGE 47
              HIP OR VALLEY   – SEAT   33.5
                "   "   "   – RIDGE 56.5
              JACK RAFTER   – EDGE   36
              PURLIN   – EDGE   54
                "   – SIDE   55.5

JACK RAFTERS 333 mm **CENTRES DECREASE**   455 (in mm to 999 and
               400 "   "   "   547 thereafter in m)
               500 "   "   "   684
               600 "   "   "   820

Run of Rafter	0.1	0.2	0.3	0.4	0.5	0.6	0.7	0.8	0.9	1.0
Length of Rafter	0.137	0.273	0.41	0.547	0.684	0.82	0.967	1.094	1.231	1.367
Length of Hip	0.169	0.339	0.508	0.678	0.847	1.016	1.186	1.355	1.525	1.694

RISE OF COMMON RAFTER 0.966 m PER METRE OF RUN   PITCH 44°

BEVELS:	COMMON RAFTER	– SEAT	44
	" "	– RIDGE	46
	HIP OR VALLEY	– SEAT	34.5
	" " "	– RIDGE	55.5
	JACK RAFTER	– EDGE	35.5
	PURLIN	– EDGE	54.5
	"	– SIDE	55

JACK RAFTERS 333 mm **CENTRES DECREASE** 463 (in mm to 999 and
400 "   "   "   556   thereafter in m)
500 "   "   "   695
600 "   "   "   834

Run of Rafter	0.1	0.2	0.3	0.4	0.5	0.6	0.7	0.8	0.9	1.0
Length of Rafter	0.139	0.278	0.417	0.556	0.695	0.834	0.973	1.111	1.251	1.39
Length of Hip	0.171	0.342	0.514	0.685	0.856	1.027	1.199	1.37	1.541	1.712

RISE OF COMMON RAFTER 1.0 m **PER METRE OF RUN**   **PITCH** 45°

BEVELS: COMMON RAFTER   – SEAT   45
             ″   ″   – RIDGE 45
          HIP OR VALLEY   – SEAT   35.5
           ″   ″   ″   – RIDGE 54.5
          JACK RAFTER   – EDGE   35.5
          PURLIN   – EDGE   54.5
             ″   – SIDE   54.5

JACK RAFTERS   333 mm   **CENTRES DECREASE**   471   (in mm to 999 and
                   400 ″   ″   ″   566   thereafter in m)
                   500 ″   ″   ″   707
                   600 ″   ″   ″   848

Run of Rafter	0.1	0.2	0.3	0.4	0.5	0.6	0.7	0.8	0.9	1.0
Length of Rafter	0.141	0.283	0.424	0.566	0.707	0.848	0.99	1.131	1.273	1.414
Length of Hip	0.173	0.346	0.519	0.693	0.866	1.039	1.212	1.386	1.559	1.732

RISE OF COMMON RAFTER 1.036 m **PER METRE OF RUN**   **PITCH** 46°

BEVELS: COMMON RAFTER   – SEAT 46
                   "   "   – RIDGE 44
             HIP OR VALLEY   – SEAT 36
               "   "   "   – RIDGE 54
             JACK RAFTER   – EDGE 35
             PURLIN   – EDGE 55
                   "   – SIDE 54.5

JACK RAFTERS 333 mm **CENTRES DECREASE**   480   (in mm to 999 and
                        400 "   "   "   576   thereafter in m)
                        500 "   "   "   720
                        600 "   "   "   864

Run of Rafter	0.1	0.2	0.3	0.4	0.5	0.6	0.7	0.8	0.9	1.0
Length of Rafter	0.144	0.288	0.432	0.576	0.72	0.864	1.058	1.152	1.296	1.44
Length of Hip	0.175	0.351	0.526	0.701	0.876	1.052	1.227	1.402	1.578	1.753

RISE OF COMMON RAFTER 1.072 m **PER METRE OF RUN**   **PITCH**   47°

BEVELS:	COMMON RAFTER	– SEAT	47
	" "	– RIDGE	43
	HIP OR VALLEY	– SEAT	37
	" " "	– RIDGE	53
	JACK RAFTER	– EDGE	34.5
	PURLIN	– EDGE	55.5
	"	– SIDE	54

JACK RAFTERS 333 mm **CENTRES DECREASE**   488   (in mm to 999 and
                 400 "   "   "   586   thereafter in m)
                 500 "   "   "   733
                 600 "   "   "   880

Run of Rafter	0.1	0.2	0.3	0.4	0.5	0.6	0.7	0.8	0.9	1.0
Length of Rafter	0.147	0.293	0.44	0.587	0.733	0.88	1.026	1.173	1.32	1.466
Length of Hip	0.177	0.355	0.532	0.71	0.887	1.065	1.242	1.42	1.597	1.775

RISE OF COMMON RAFTER 1.111 m PER METRE OF RUN   PITCH 48°

BEVELS:	COMMON RAFTER	– SEAT	48
	" "	– RIDGE	42
	HIP OR VALLEY	– SEAT	38
	" " "	– RIDGE	52
	JACK RAFTER	– EDGE	34
	PURLIN	– EDGE	56
	"	– SIDE	53.5

JACK RAFTERS 333 mm **CENTRES DECREASE** 498 (in mm to 999 and
                    400 "   "   "   598 thereafter in m)
                    500 "   "   "   747
                    600 "   "   "   896

Run of Rafter	0.1	0.2	0.3	0.4	0.5	0.6	0.7	0.8	0.9	1.0
Length of Rafter	0.149	0.299	0.448	0.598	0.747	0.897	1.046	1.196	1.345	1.494
Length of Hip	0.18	0.36	0.539	0.719	0.899	1.079	1.259	1.438	1.618	1.798

RISE OF COMMON RAFTER 1.15 m **PER METRE OF RUN**   **PITCH** 49°

BEVELS: COMMON RAFTER – SEAT 49
                         ″   ″ – RIDGE 41
              HIP OR VALLEY – SEAT 39
                ″   ″   ″ – RIDGE 51
              JACK RAFTER – EDGE 33.5
              PURLIN – EDGE 56.5
                ″ – SIDE 53

JACK RAFTERS 333 mm **CENTRES DECREASE** 508 (in mm to 999 and
                     400 ″   ″   ″ 610 thereafter in m)
                     500 ″   ″   ″ 762
                     600 ″   ″   ″ 914

Run of Rafter	0.1	0.2	0.3	0.4	0.5	0.6	0.7	0.8	0.9	1.0
Length of Rafter	0.152	0.305	0.457	0.61	0.762	0.915	1.067	1.219	1.372	1.524
Length of Hip	0.182	0.365	0.547	0.729	0.912	1.094	1.276	1.458	1.641	1.823

RISE OF COMMON RAFTER 1.192 m **PER METRE OF RUN**   **PITCH** 50°

BEVELS: COMMON RAFTER   – SEAT   50
                 ″   ″   – RIDGE 40
            HIP OR VALLEY   – SEAT   40
             ″   ″   ″   – RIDGE 50
            JACK RAFTER   – EDGE 32.5
            PURLIN   – EDGE 57.5
                 ″   – SIDE   52.5

JACK RAFTERS   333 mm **CENTRES DECREASE**   518   (in mm to 999 and
                   400 ″   ″   ″   622   thereafter in m)
                   500 ″   ″   ″   778
                   600 ″   ″   ″   934

Run of Rafter	0.1	0.2	0.3	0.4	0.5	0.6	0.7	0.8	0.9	1.0
Length of Rafter	0.156	0.311	0.467	0.622	0.778	0.933	1.089	1.246	1.4	1.556
Length of Hip	0.185	0.37	0.555	0.74	0.925	1.11	1.295	1.48	1.664	1.849

RISE OF COMMON RAFTER 1.235 m **PER METRE OF RUN**   **PITCH** 51°

BEVELS: COMMON RAFTER – SEAT 51
                                  ″   ″   – RIDGE 39
                  HIP OR VALLEY – SEAT 41
                  ″   ″   ″   – RIDGE 49
                  JACK RAFTER – EDGE 32
                  PURLIN – EDGE 58
                          ″   – SIDE 52

JACK RAFTERS 333 mm **CENTRES DECREASE** 529 (in mm to 999 and
                     400 ″   ″   ″   636 thereafter in m)
                     500 ″   ″   ″   795
                     600 ″   ″   ″   953

Run of Rafter	0.1	0.2	0.3	0.4	0.5	0.6	0.7	0.8	0.9	1.0
Length of Rafter	0.159	0.318	0.477	0.636	0.795	0.953	1.012	1.271	1.43	1.589
Length of Hip	0.188	0.375	0.563	0.751	0.936	1.126	1.314	1.502	1.69	1.877

RISE OF COMMON RAFTER 1.28 m **PER METRE OF RUN** **PITCH** 52°

BEVELS: COMMON RAFTER – SEAT 52
          "   " – RIDGE 38
          HIP OR VALLEY – SEAT 42
          "   "   " – RIDGE 48
          JACK RAFTER – EDGE 31.5
          PURLIN – EDGE 58.5
          " – SIDE 52

JACK RAFTERS 333 mm **CENTRES DECREASE** 541 (in mm to 999 and
                400 "   "   "   650 thereafter in m)
                500 "   "   "   812
                600 "   "   "   974

Run of Rafter	0.1	0.2	0.3	0.4	0.5	0.6	0.7	0.8	0.9	1.0
Length of Rafter	0.162	0.325	0.487	0.65	0.812	0.974	1.137	1.299	1.462	1.624
Length of Hip	0.191	0.381	0.572	0.763	0.954	1.144	1.335	1.526	1.717	1.907

RISE OF COMMON RAFTER 1.327 m **PER METRE OF RUN**   **PITCH** 53°

BEVELS: COMMON RAFTER   – SEAT   53
                    ″   ″   – RIDGE 37
                 HIP OR VALLEY   – SEAT   43
                 ″   ″   ″   – RIDGE 47
                 JACK RAFTER   – EDGE   31
                 PURLIN   – EDGE   59
                      ″   – SIDE   51.5

JACK RAFTERS 333 mm **CENTRES DECREASE**   553   (in mm to 999 and
                  400 ″   ″   ″   665   thereafter in m)
                  500 ″   ″   ″   831
                  600 ″   ″   ″   997

Run of Rafter	0.1	0.2	0.3	0.4	0.5	0.6	0.7	0.8	0.9	1.0
Length of Rafter	0.166	0.332	0.498	0.665	0.831	0.997	1.163	1.329	1.495	1.662
Length of Hip	0.194	0.388	0.582	0.776	0.97	1.164	1.358	1.551	1.745	1.939

Calculating the length and cutting angles of timber members: data tables 5°–75°

RISE OF COMMON RAFTER   1.376 m   **PER METRE OF RUN**   **PITCH**   54°

BEVELS: COMMON RAFTER   – SEAT   54
               "   "   – RIDGE   36
        HIP OR VALLEY   – SEAT   44
        " " "   – RIDGE   46
        JACK RAFTER   – EDGE   30.5
        PURLIN   – EDGE   59.5
        "   – SIDE   51

JACK RAFTERS   333 mm   **CENTRES DECREASE**   567   (in mm to 999 and
        400 "   "   "   680   thereafter in m)
        500 "   "   "   850
        600 "   "   "   1.021

	0.1	0.2	0.3	0.4	0.5	0.6	0.7	0.8	0.9	1.0
Run of Rafter	0.1	0.2	0.3	0.4	0.5	0.6	0.7	0.8	0.9	1.0
Length of Rafter	0.17	0.34	0.51	0.681	0.851	1.021	1.191	1.361	1.531	1.701
Length of Hip	0.197	0.395	0.592	0.789	0.987	1.184	1.381	1.579	1.776	1.973

RISE OF COMMON RAFTER   1.428 m   **PER METRE OF RUN**   **PITCH**   55°

BEVELS: COMMON RAFTER   – SEAT   55
                 "   "   – RIDGE   35
            HIP OR VALLEY   – SEAT   45.5
            "   "   "   – RIDGE   44.5
            JACK RAFTER   – EDGE   30
            PURLIN   – EDGE   60
            "   – SIDE   50.5

JACK RAFTERS   333 mm   **CENTRES DECREASE**   580   (in mm to 999 and
                400 "   "   "   697   thereafter in m)
                500 "   "   "   872
                600 "   "   "   1.046

Run of Rafter	0.1	0.2	0.3	0.4	0.5	0.6	0.7	0.8	0.9	1.0
Length of Rafter	0.174	0.349	0.523	0.697	0.872	1.046	1.22	1.395	1.569	1.743
Length of Hip	0.201	0.402	0.603	0.804	1.005	1.206	1.407	1.608	1.809	2.01

RISE OF COMMON RAFTER   1.5 m   **PER METRE OF RUN**   **PITCH**   56° 18′
(Italian pitch)

BEVELS:	COMMON RAFTER	– SEAT	56.5
	" "	– RIDGE	33.5
	HIP OR VALLEY	– SEAT	46.5
	" " "	– RIDGE	43.5
	JACK RAFTER	– EDGE	29
	PURLIN	– EDGE	61
	"	– SIDE	50

JACK RAFTERS   333 mm   **CENTRES DECREASE**   600   (in mm to 999 and
                   400   "   "   720   thereafter in m)
                   500   "   "   900
                   600   "   "   1.082

Run of Rafter	0.1	0.2	0.3	0.4	0.5	0.6	0.7	0.8	0.9	1.0
Length of Rafter	0.18	0.361	0.541	0.721	0.902	1.082	1.262	1.442	1.623	1.803
Length of Hip	0.206	0.412	0.618	0.824	1.031	1.237	1.443	1.649	1.855	2.061

RISE OF COMMON RAFTER 1.6 m   **PER METRE OF RUN**   **PITCH** 58°

BEVELS: COMMON RAFTER – SEAT 58
  "   "   – RIDGE 32
  HIP OR VALLEY   – SEAT 48.5
  "   "   "   – RIDGE 41.5
  JACK RAFTER   – EDGE 28
  PURLIN   – EDGE 62
  "   – SIDE 49.5

JACK RAFTERS 333 mm **CENTRES DECREASE**   628   (in mm to 999 and
  400 "   "   "   755   thereafter in m)
  500 "   "   "   944
  600 "   "   "   1.132

Run of Rafter	0.1	0.2	0.3	0.4	0.5	0.6	0.7	0.8	0.9	1.0
Length of Rafter	0.189	0.377	0.566	0.755	0.944	1.132	1.321	1.51	1.698	1.887
Length of Hip	0.214	0.428	0.641	0.855	1.069	1.283	1.496	1.71	1.924	2.136

RISE OF COMMON RAFTER   1.664 m   **PER METRE OF RUN**   **PITCH**   59°

BEVELS: COMMON RAFTER   – SEAT   59
                "   "   – RIDGE   31
          HIP OR VALLEY   – SEAT   49.5
             "   "   "   – RIDGE   40.5
          JACK RAFTER   – EDGE   27.5
          PURLIN   – EDGE   62.5
               "   – SIDE   49.5

JACK RAFTERS   333 mm   **CENTRES DECREASE**   647   (in mm to 999 and
                   400 "   "   "   777   thereafter in m)
                   500 "   "   "   971
                   600 "   "   "   1.165

Run of Rafter	0.1	0.2	0.3	0.4	0.5	0.6	0.7	0.8	0.9	1.0
Length of Rafter	0.194	0.398	0.582	0.777	0.971	1.165	1.359	1.553	1.747	1.942
Length of Hip	0.218	0.437	0.655	0.874	1.092	1.31	1.529	1.747	1.965	2.184

RISE OF COMMON RAFTER   1.732 m   **PER METRE OF RUN**   **PITCH** 60°
(Equilateral pitch)

BEVELS: COMMON RAFTER   – SEAT   60
        "   "   – RIDGE   30
     HIP OR VALLEY   – SEAT   51
      "   "   "   – RIDGE   39
     JACK RAFTER   – EDGE   26.5
     PURLIN   – EDGE   63.5
        "   – SIDE   49

JACK RAFTERS   333 mm   **CENTRES DECREASE**   666   (in mm to 999 and
        400   "   "   "   800   thereafter in m)
        500   "   "   "   1.000
        600   "   "   "   1.200

Run of Rafter	0.1	0.2	0.3	0.4	0.5	0.6	0.7	0.8	0.9	1.0
Length of Rafter	0.2	0.4	0.6	0.8	1.0	1.2	1.4	1.6	1.8	2.0
Length of Hip	0.224	0.447	0.671	0.894	1.118	1.342	1.565	1.789	2.012	2.236

RISE OF COMMON RAFTER   1.804 m   **PER METRE OF RUN**   **PITCH**   61°

BEVELS:	COMMON RAFTER	– SEAT	61
	" "	– RIDGE	29
	HIP OR VALLEY	– SEAT	52
	" " "	– RIDGE	38
	JACK RAFTER	– EDGE	26
	PURLIN	– EDGE	64
	"	– SIDE	49

JACK RAFTERS   333 mm   **CENTRES DECREASE**   687   (in mm to 999 and
                   400 "   "   "   825   thereafter in m)
                   500 "   "   "   1.032
                   600 "   "   "   1.238

Run of Rafter	0.1	0.2	0.3	0.4	0.5	0.6	0.7	0.8	0.9	1.0
Length of Rafter	0.206	0.413	0.619	0.825	1.031	1.238	1.444	1.65	1.857	2.063
Length of Hip	0.229	0.458	0.688	0.917	1.146	1.375	1.605	1.834	2.063	2.292

RISE OF COMMON RAFTER   1.88 m   **PER METRE OF RUN**   **PITCH**   62°

BEVELS: COMMON RAFTER   – SEAT   62
          ″   ″   – RIDGE 28
          HIP OR VALLEY   – SEAT   53
          ″   ″   ″   – RIDGE 37
          JACK RAFTER   – EDGE   25
          PURLIN   – EDGE   65
          ″   – SIDE   48.5

JACK RAFTERS 333 mm **CENTRES DECREASE**   709   (in mm to 999 and
                  400 ″   ″   ″   852   thereafter in m)
                  500 ″   ″   ″   1.065
                  600 ″   ″   ″   1.278

Run of Rafter	0.1	0.2	0.3	0.4	0.5	0.6	0.7	0.8	0.9	1.0
Length of Rafter	0.213	0.426	0.639	0.852	1.065	1.278	1.491	1.704	1.917	2.13
Length of Hip	0.235	0.471	0.706	0.941	1.177	1.412	1.647	1.882	2.118	2.353

RISE OF COMMON RAFTER 2.0 m **PER METRE OF RUN** **PITCH** 63° 26′
(Gothic pitch)

BEVELS:	COMMON RAFTER	– SEAT	63.5
	" "	– RIDGE	26.5
	HIP OR VALLEY	– SEAT	54.5
	" " "	– RIDGE	35.5
	JACK RAFTER	– EDGE	24
	PURLIN	– EDGE	66
	"	– SIDE	48

JACK RAFTERS 333 mm **CENTRES DECREASE** 745 (in mm to 999 and
400 " " " 894 thereafter in m)
500 " " " 1.118
600 " " " 1.342

Run of Rafter	0.1	0.2	0.3	0.4	0.5	0.6	0.7	0.8	0.9	1.0
Length of Rafter	0.224	0.447	0.671	0.894	1.118	1.342	1.565	1.789	2.012	2.236
Length of Hip	0.245	0.49	0.735	0.98	1.225	1.47	1.715	1.96	2.205	2.45

RISE OF COMMON RAFTER 2.145 m **PER METRE OF RUN**   **PITCH** 65°

BEVELS: COMMON RAFTER – SEAT 65
               "   " – RIDGE 25
         HIP OR VALLEY – SEAT 56.5
            "   "   " – RIDGE 33.5
         JACK RAFTER – EDGE 23
         PURLIN – EDGE 67
             " – SIDE 48

JACK RAFTERS 333 mm **CENTRES DECREASE** 788 (in mm to 999 and
             400 "   "   "   946 thereafter in m)
             500 "   "   "   1.183
             600 "   "   "   1.420

Run of Rafter	0.1	0.2	0.3	0.4	0.5	0.6	0.7	0.8	0.9	1.0
Length of Rafter	0.237	0.473	0.71	0.946	1.183	1.42	1.656	1.893	2.13	2.366
Length of Hip	0.257	0.514	0.771	1.028	1.284	1.541	1.798	2.055	2.312	2.569

RISE OF COMMON RAFTER 2.246 m   **PER METRE OF RUN**   **PITCH** 66°

BEVELS: COMMON RAFTER   – SEAT 66
               ″   ″   – RIDGE 24
          HIP OR VALLEY   – SEAT 58
            ″ ″ ″   – RIDGE 32
          JACK RAFTER   – EDGE 22
          PURLIN   – EDGE 68
            ″   – SIDE 47.5

JACK RAFTERS 333 mm **CENTRES DECREASE**   819   (in mm to 999 and
                  400 ″   ″   ″   984   thereafter in m)
                  500 ″   ″   ″   1.230
                  600 ″   ″   ″   1.475

Run of Rafter	0.1	0.2	0.3	0.4	0.5	0.6	0.7	0.8	0.9	1.0
Length of Rafter	0.246	0.492	0.738	0.983	1.229	1.475	1.721	1.967	2.213	2.459
Length of Hip	0.265	0.531	0.796	1.062	1.327	1.593	1.858	2.123	2.389	3.654

RISE OF COMMON RAFTER 2.356 m **PER METRE OF RUN** **PITCH** 67°

BEVELS: COMMON RAFTER – SEAT 67
" " – RIDGE 23
HIP OR VALLEY – SEAT 59
" " " – RIDGE 31
JACK RAFTER – EDGE 21.5
PURLIN – EDGE 68.5
" – SIDE 47.5

JACK RAFTERS 333 mm **CENTRES DECREASE** 852 (in mm to 999 and
400 " " " 1.024 thereafter in m)
500 " " " 1.280
600 " " " 1.535

Run of Rafter	0.1	0.2	0.3	0.4	0.5	0.6	0.7	0.8	0.9	1.0
Length of Rafter	0.256	0.512	0.768	1.024	1.28	1.536	1.792	2.047	2.303	2.559
Length of Hip	0.275	0.55	0.824	1.099	1.374	1.649	1.922	2.198	2.473	2.748

RISE OF COMMON RAFTER 2.475 m **PER METRE OF RUN**   **PITCH** 68°

BEVELS: COMMON RAFTER   – SEAT 68
                   ″   ″   – RIDGE 22
          HIP OR VALLEY   – SEAT   60.5
          ″   ″   ″   – RIDGE 29.5
          JACK RAFTER   – EDGE   20.5
          PURLIN   – EDGE   69.5
          ″   – SIDE   47

JACK RAFTERS 333 mm **CENTRES DECREASE**   889   (in mm to 999 and
             400 ″   ″   ″   1.068   thereafter in m)
             500 ″   ″   ″   1.335
             600 ″   ″   ″   1.601

Run of Rafter	0.1	0.2	0.3	0.4	0.5	0.6	0.7	0.8	0.9	1.0
Length of Rafter	0.267	0.534	0.801	1.068	1.335	1.602	1.869	2.136	2.403	2.669
Length of Hip	0.285	0.57	0.855	1.14	1.425	1.71	1.995	2.28	2.566	2.851

RISE OF COMMON RAFTER 2.605 m   **PER METRE OF RUN**   **PITCH** 69°

BEVELS: COMMON RAFTER   – SEAT 60
          "   "   – RIDGE 21
          HIP OR VALLEY   – SEAT 61.5
          "   "   "   – RIDGE 28.5
          JACK RAFTER   – EDGE 19.5
          PURLIN   – EDGE 70.5
          "   – SIDE 47

JACK RAFTERS 333 mm **CENTRES DECREASE**   930   (in mm to 999 and
                  400 "   "   "   1.116 thereafter in m)
                  500 "   "   "   1.395
                  600 "   "   "   1.674

	0.1	0.2	0.3	0.4	0.5	0.6	0.7	0.8	0.9	1.0
Run of Rafter	0.1	0.2	0.3	0.4	0.5	0.6	0.7	0.8	0.9	1.0
Length of Rafter	0.279	0.558	0.837	1.116	1.395	1.674	1.953	2.232	2.511	2.79
Length of Hip	0.296	0.593	0.889	1.186	1.482	1.779	2.075	2.371	2.688	2.964

RISE OF COMMON RAFTER 2.747 m   **PER METRE OF RUN**   **PITCH** 70°

BEVELS: COMMON RAFTER   – SEAT   70
       "   "   – RIDGE 20
      HIP OR VALLEY   – SEAT   63
      " " "   – RIDGE 27
      JACK RAFTER   – EDGE   19
      PURLIN   – EDGE   71
      "   – SIDE   47

JACK RAFTERS 333 mm **CENTRES DECREASE**   975   (in mm to 999 and
       400 "   "   "   1.170   thereafter in m)
       500 "   "   "   1.462
       600 "   "   "   1.754

Run of Rafter	0.1	0.2	0.3	0.4	0.5	0.6	0.7	0.8	0.9	1.0
Length of Rafter	0.292	0.585	0.877	1.17	1.462	1.754	2.047	2.339	2.631	2.924
Length of Hip	0.309	0.618	0.927	1.236	1.545	1.854	2.163	2.472	2.781	3.09

RISE OF COMMON RAFTER 2.904 m **PER METRE OF RUN**   **PITCH** 71°

BEVELS: COMMON RAFTER   – SEAT 71
         "   "   – RIDGE 19
         HIP OR VALLEY   – SEAT 64
         "   "   "   – RIDGE 26
         JACK RAFTER   – EDGE 18
         PURLIN   – EDGE 72
         "   – SIDE 46.5

JACK RAFTERS 333 mm **CENTRES DECREASE** 1.024 (in mm to 999 and
                  400 "   "   "   1.229   thereafter in m)
                  500 "   "   "   1.536
                  600 "   "   "   1.843

Run of Rafter	0.1	0.2	0.3	0.4	0.5	0.6	0.7	0.8	0.9	1.0
Length of Rafter	0.307	0.614	0.921	1.229	1.536	1.843	2.15	2.457	2.764	3.072
Length of Hip	0.323	0.646	0.969	1.292	1.615	1.938	2.261	2.584	2.907	3.23

RISE OF COMMON RAFTER   3.078 m   **PER METRE OF RUN**   **PITCH**   72°

BEVELS:	COMMON RAFTER	– SEAT	72
	" "	– RIDGE	18
	HIP OR VALLEY	– SEAT	65.5
	" " "	– RIDGE	24.5
	JACK RAFTER	– EDGE	17
	PURLIN	– EDGE	73
	"	– SIDE	46.5

JACK RAFTERS   333 mm   **CENTRES DECREASE**   1.078   (in mm to 999 and
                  400 "   "   "   1.294   thereafter in m)
                  500 "   "   "   1.618
                  600 "   "   "   1.942

Run of Rafter	0.1	0.2	0.3	0.4	0.5	0.6	0.7	0.8	0.9	1.0
Length of Rafter	0.324	0.647	0.971	1.294	1.618	1.942	2.266	2.589	2.912	3.236
Length of Hip	0.339	0.677	1.016	1.355	1.694	2.032	2.371	2.71	3.048	3.387

RISE OF COMMON RAFTER   3.271 m **PER METRE OF RUN**   **PITCH**   73°

BEVELS: COMMON RAFTER   – SEAT   73
               "   "   – RIDGE   17
          HIP OR VALLEY   – SEAT   66.5
               "   "   "   – RIDGE   23.5
          JACK RAFTER   – EDGE   16.5
          PURLIN   – EDGE   73.5
               "   – SIDE   46.5

JACK RAFTERS   333 mm   **CENTRES DECREASE**   1.140   (in mm to 999 and
                   400 "   "   "   1.368   thereafter in m)
                   500 "   "   "   1.710
                   600 "   "   "   2.052

Run of Rafter	0.1	0.2	0.3	0.4	0.5	0.6	0.7	0.8	0.9	1.0
Length of Rafter	0.342	0.684	1.026	1.368	1.71	2.052	2.394	2.736	3.078	3.42
Length of Hip	0.356	0.713	1.069	1.425	1.782	2.138	2.494	2.851	3.207	3.563

Calculating the length and cutting angles of timber members: data tables 5°–75°

RISE OF COMMON RAFTER 3.487 m **PER METRE OF RUN**   **PITCH** 74°

BEVELS:	COMMON RAFTER	– SEAT	74
	" "	– RIDGE	16
	HIP OR VALLEY	– SEAT	68
	" " "	– RIDGE	22
	JACK RAFTER	– EDGE	15.5
	PURLIN	– EDGE	74.5
	"	– SIDE	46

JACK RAFTERS 333 mm **CENTRES DECREASE** 1.209 (in mm to 999 and
    400 "   "   "   1.451   thereafter in m)
    500 "   "   "   1.814
    600 "   "   "   2.177

Run of Rafter	0.1	0.2	0.3	0.4	0.5	0.6	0.7	0.8	0.9	1.0
Length of Rafter	0.363	0.726	1.088	1.451	1.814	2.177	2.54	2.902	3.265	3.628
Length of Hip	0.376	0.753	1.129	1.505	1.882	2.258	2.634	3.011	3.387	3.763

RISE OF COMMON RAFTER   3.732 m   **PER METRE OF RUN**   **PITCH**   75°

BEVELS:	COMMON RAFTER	– SEAT	75
	" "	– RIDGE	15
	HIP OR VALLEY	– SEAT	69
	" " "	– RIDGE	21
	JACK RAFTER	– EDGE	14.5
	PURLIN	– EDGE	75.5
	"	– SIDE	46

JACK RAFTERS 333 mm **CENTRES DECREASE**   1.289   (in mm to 999 and
                  400 "   "   "   1.546   thereafter in m)
                  500 "   "   "   1.932
                  600 "   "   "   2.318

Run of Rafter	0.1	0.2	0.3	0.4	0.5	0.6	0.7	0.8	0.9	1.0
Length of Rafter	0.386	0.773	1.159	1.545	1.932	2.318	2.705	3.091	3.477	3.864
Length of Hip	0.399	0.789	1.197	1.596	1.996	2.395	2.794	3.193	3.592	3.991

5   WALL PLATES – STRAPPING AND GABLE STRAPPING

Earlier in this book it was stated that the wall plate is the foundation to the roof, and like all foundations needs to be sound and secure. The wall plate should be strapped down to the building structure; this is usually done by using steel straps, as illustrated in Figure 5.1(a). The straps can either be built into the brick wall below, or face-fixed by screwing into the wall. On timber-framed housing the plate may be adequately secured to the frame by nailing at centres specified by the designer.

Gable walls – especially those on steep pitched roof constructions where the gable is very tall – rely on the roof for their stability and NOT the other way around. Wind blowing on one gable exerts a pressure on it pushing it into the roof, whilst at the opposite end it creates a suction which attempts to suck the gable from the roof. It is therefore a requirement of the building regulations that the gables must be adequately tied back into the roof to give them support. Figure 5.1(b) illustrates a typical gable end restraint system on a trussed rafter roof, but this equally applies to traditional roof construction. Building the purlins into the gable will help, but will only be effective if the purlin is mechanically fixed to the wall with some additional form of cleat or strap. Straps are normally placed at approximately 2 m centres, and it is essential that the strap is supported by solid blocking beneath it to ensure that it does not buckle, and that the last rafter is solidly blocked to the gable wall itself.

Goss's Roofing Ready Reckoner: From Timberwork to Tiles, Fifth Edition. C. N. Mindham.
© 2016 John Wiley & Sons, Ltd. Published 2016 by John Wiley & Sons, Ltd.

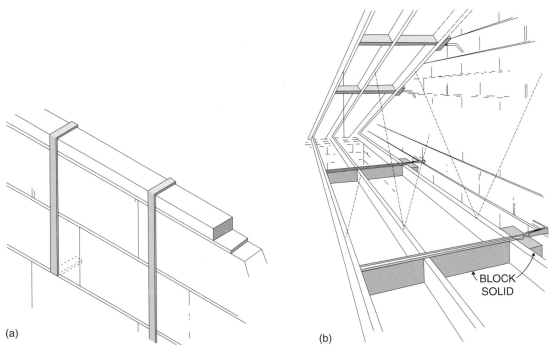

(a)

(b)

BLOCK
SOLID

Figure 5.1 Wall plate and Gable Wall strapping.

6 WIND BRACING AND OPENINGS FOR DORMERS AND ROOF WINDOWS

From the previous chapter it can be seen that wind plays a considerable part in destabilising a structure, and measures must be taken to ensure the stability of the roof construction in high wind situations. From the previous chapter, strapping the gables to the roof has ensured their integrity with the roof, but apart from the binders, purlins and ridge – which connect the rafter members on a horizontal plane – the roof structure is still no more than a number of vertical members of timber connected with a limited number of nails to the members mentioned above. The roof is in effect no more than a set of dominoes standing vertically on their ends and which can be easily made to fall if a pressure is applied to the last one in line; in the case of a roof this would be the gable end. To prevent the domino toppling effect, wind bracing is introduced into the structure to triangulate it on a vertical plane. The elements being discussed here are set out in Figure 6.1, which details the bracing required for a typical trussed rafter roof, much of this being applicable to a traditionally constructed roof without hips. Even on a hip roof where the ridge is twice as long as the length on plan of the hip itself, it is wise to introduce wind bracing.

Wind bracing is usually timber typically of 25 × 100 mm in cross-section, fitted from the wall plate to the ridge at an angle of approximately 45° on the underside of the rafters. At each rafter

Goss's Roofing Ready Reckoner: From Timberwork to Tiles, Fifth Edition. C. N. Mindham.
© 2016 John Wiley & Sons, Ltd. Published 2016 by John Wiley & Sons, Ltd.

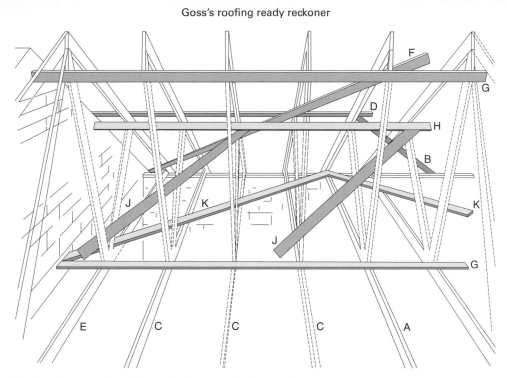

Figure 6.1 Bracing required for typical trussed rafter roof.

crossing, the wind bracing should be nailed to the rafter with three nails. The braces should be fitted from the foot of the gable to the ridge on both sides of the roof, and then at 45° back down again to the plate over the entire length of the roof. This is brace F in Figure 6.1; brace H may well be replaced by a purlin in a traditionally cut roof, and brace G by the ridge board. Brace G at ceiling joist level, may well be ceiling joist binders on a traditionally cut roof, whilst K should be fitted to all types of roof construction. Brace J applies only to trussed rafter roof construction.

OPENINGS FOR DORMERS AND ROOF WINDOWS

If part or the whole of the roof is to be used as a living space, or simply a hobby room, or perhaps some natural light is required in the roof void for some other purpose, then openings may have to be formed in the roof structure. The following text gives some guidance on openings in newly built traditional cut roofs, conversion work in traditional cut roofs, new trussed rafter roofs, and existing trussed rafter roofs.

Traditional Cut Roofs (not Trussed Rafter Construction)

New Build Roofs
Openings for dormers and roof windows may be formed by adding timbers to the sides of the opening as set out in Figure 6.2(a) and (b) (these are taken from *Roof Construction and Loft Conversion*, by the present author). It is strongly recommended that all timber connections – that is, trimmed rafter to trimmer and trimmer to double or triple rafter – are made with metal connection plates, fixed using the correct nails or screws. This will give a stronger joint than simply attempting to angular or 'tosh' nail. Whilst 'tosh' nailing is a traditional way of connecting roof timbers, it is very prone to splitting the timber, resulting in a poor joint.

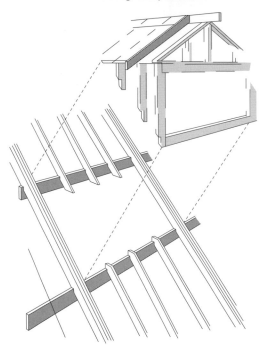

Figure 6.2a Trimmed opening for dormer. Reproduced from C. N. Mindham (2006) *Roof Construction and Loft Conversion*, 4th edn. Blackwell, UK.

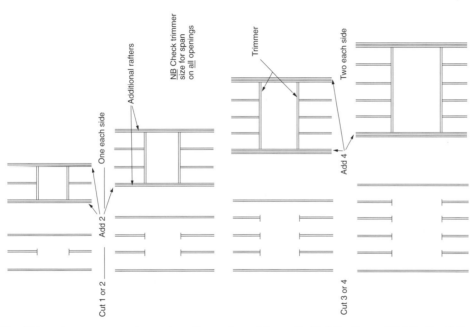

Figure 6.2b Trimming construction rules. Reproduced from C. N. Mindham (2006) *Roof Construction and Loft Conversion*, 4th edn. Blackwell, UK.

Conversion Work

The same rules as set out above will apply, but additionally a purlin may have to be removed to give adequate opening for the new dormer or roof window. In this case a further purlin should be inserted above and below the opening and securely fixed to its supporting walls or purlin post. The existing roof should then be fitted to it BEFORE removing the old purlin and cutting the rafters away to form the roof opening. Additional rafters either side of the opening may not be required if the purlins are immediately above and below the opening itself. Trimmers should still be fitted to pick up the ends of the rafters to completely frame the opening. Possible roof window installations are shown in Figure 6.3.

Trussed Rafter Roofs

New Roofs

The trussed rafter roof designer will take any openings in the roof, including those for chimneys, into account in the calculation of the whole roof design. Specific rules apply to forming openings in trussed rafter roofs, and on NO ACCOUNT should the rules given above for traditional roofs be applied to trussed rafter constructions. Trussed rafter roofs are generally now a 'whole roof' design, and not simply the supply of trussed rafters to be erected as the builder wishes. Make use of this design service, which is available through your trussed rafter supplier, to ensure a sound roof structure.

Existing Trussed Rafter Roofs

Trussed rafters should NEVER be cut, as this will endanger the whole roof structure. Consult the original roof designer or manufacturer if known, and seek their advice on any modifications to form openings in the roof that may be proposed. Details of the original manufacturer should be found on a card fixed under the nail plate at the ridge of the trussed rafter; this will give details of

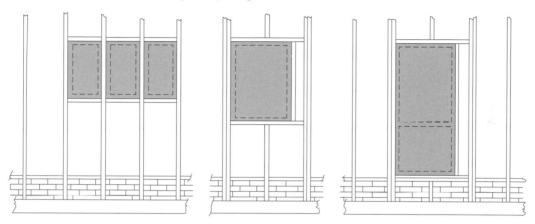

Figure 6.3 Some typical roof window installations. Reproduced from C. N. Mindham (2006) *Roof Construction and Loft Conversion*, 4th edn. Blackwell, UK, p. 216.

the manufacturer as well as the nail plate system provider. Until fairly recently the identity card was not standardised as being fixed at the ridge, and it may therefore be fixed under any other plate on the structure. Early trussed rafters will have no form of identification of their manufacturer. In that case there is no alternative but to consult a structural engineer (see Fig. 8.1).

However, as most trussed rafter roofs have trussed rafters spaced at 600 mm centres, it is possible to buy roof windows that are designed specifically to fit neatly between the trussed rafters, thus avoiding any structural modifications to the roof structure. It is possible to couple a number of these windows side-by-side along the length of the roof, giving the effect of a much larger single opening. This clearly avoids the expense of altering the roof structure, and is much more simple and cost effective. As this approach results in a multiple number of small windows, it affords more ventilation options than one large window. The left-hand illustration in Figure 6.3 indicates a multiple roof window installation in a trussed rafter or, indeed, in a traditionally constructed cut roof.

7   ROOFING METALWORK AND FIXINGS

The use of smaller timber sections, particularly in engineered roofs such as trussed rafter pre-fabricated assemblies, has led to the increased use of metalwork to join the various members together. Traditional 'tosh' or 'skew' nailing can easily split these smaller timber sections, resulting in a poor connection. Figure 7.1 illustrates some items suitable for cut roof and trussed rafter construction. Today, with the increased use of 'I' beams, additional special metalwork is available, specifically for that product, and also for metal web beams and glued laminated timber beams (glulam). One company which produces a very comprehensive range of metalwork timber construction is SIMPSON Strong-Tie (for details, see Bibliography). (An illustration of special metalwork designed for use with timber 'I' beams, when used as floor joists for attic rooms, is shown in Figure 8.5.)

All metalwork used in roof construction should be galvanized (this includes the nails) to prevent corrosion. The modern roof can suffer from localised condensation, especially if not adequately ventilated, and this could lead to the premature corrosion of nails and fixings. Whilst a galvanised nail has a slightly rough surface compared to a plain wire nail, and therefore gives an improved resistance to movement in the joint, many of the metal-to-timber connections are specified to be fixed with square-twisted galvanised nails which give a far improved performance in the joint.

Goss's Roofing Ready Reckoner: From Timberwork to Tiles, Fifth Edition. C. N. Mindham.
© 2016 John Wiley & Sons, Ltd. Published 2016 by John Wiley & Sons, Ltd.

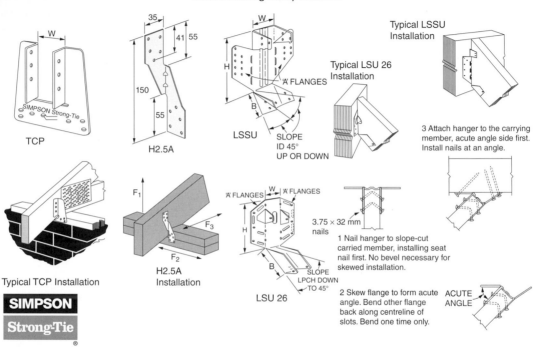

Figure 7.1 Some items of roofing metalwork. Illustration taken from Simpson Strongtie, Cat. Connections for Masonry and Timber Construction C-UK 14, p 145.

Typical roofing metalwork would be as follows:

(a) Wall plate straps cross-section 30 mm × 1.5/2 mm.
(b) Gable restraint straps 30 mm × 5 mm.
(c) Trussed rafter clips to hold truss or rafter to wall plate.
(d) Hip corner tie to hold hip to wall plate at corner.
(e) Girder truss shoes to carry trussed rafters on girder truss support points.
(f) Multi-nail plates for coupling timbers into longer lengths.
(g) Framing anchors to connect various trimmed openings and elements in the ceiling joist structure.

Care must be taken to use the nails specified by the metalwork manufacturer, both in the type of nail to be used and the number of nails to be used in each joint. (See also the TRADA publication 'Timber Engineering Hardware and Connectors'; Ref WIS 2/3-51.) For contact details see below.

NAILS, BOLTS AND SCREWS

Whilst, traditionally, roof structures have been connected using nails – and continue to be connected so – there is an ever-increasing use of light metal connectors (as shown in Fig. 7.1), complete with their specialist nails. The use of traditional large round wire nails (galvanised for roof structures) can lead to serious splitting of the timbers, thus resulting in a poor strength of joint. It is also a practised skill to be able to nail a rafter to a wall plate using 'tosh' or 'skew' nailing, and this often leads to misplacement of the rafter itself. Again, a better method is to use a framing anchor or a truss clip, as illustrated. However, with the ready availability of powerful battery-operated drills/drivers, the application of bolts and high-tensile extremely long steel screws makes the construction of a roof not only easier but more precise and definitely stronger. If bolts are to be used, then either a hexagon head or a cup square-head coach bolt could be used, but in either case a large square or round washer should be placed under both the head of the bolt and the nut to avoid crushing the timber. Washers are traditionally 50 mm × 50 mm × 3 mm thick or 50 mm or 63 mm in diameter. Washers must NOT BE OVER TIGHTENED, as this does not necessarily greatly increase the strength of the connection. The object is NOT to distort the washer by bedding it into the timber, as this could cause splitting.

The modern especially designed long timber screws are manufactured from carbon steel and are epoxy-coated to provide corrosion protection while at the same time lowering friction when driving the screw. These relatively slender screws do not require pilot holes, and are driven by a special adaptor tool provided with the screws, driven by a minimum of 18 V battery-powered drill driver. The application is extremely fast and provides a very strong joint. These screws are particularly efficient and time-saving compared to a bolt when making connections between two trimmers such as those at the side of roof window openings. Their lengths range from 66 to 135 mm. One manufacturer of such screws is Fasten Master, and they are available from many builders' merchants and timber suppliers (for further details, see Bibliography).

More detailed information on fixings for structural timber, reference could be made to a TRADA Publication entitled 'FASTENERS FOR STRUCTURAL TIMBER: Nails, Staples, Screws, Dowels and Bolts'. This is one of their 'Wood Information Sheets'. No. WIS2/3-52. Contact details for TRADA are available in the Bibliography.

8 ENGINEERED TIMBER ROOFING COMPONENTS AND PREFABRICATED ROOFS

TRUSSED RAFTERS

Trussed rafters are the most simple form of prefabricated roofing component, combining rafters and ceiling joist in one unit. The rafter and ceiling joist are supported within the unit by triangulated smaller members called 'webs'. The connections at timber joints are made by using multi-nail plates, pressed into the timbers by special machines. The timbers used are all stress graded and, being assembled in a jig, the quality of production accuracy is assured. Each trussed rafter must have a manufacturer's identification plate affixed as part of the pressing process, to prove the quality standards to which it has been produced. An example of one such identification label is shown in Figure 8.1.

Trussed rafters are designed to carry all the loads imposed on them by self weight, underlay, battens, tiles and snow, provided they are kept truly vertical when built into the roof. To ensure this, longitudinal bracing is fitted to keep them at constant spacing and to resist the buckling of some elements of the truss, and also diagonal wind bracing; the latter is extremely important to ensure a rigid whole-roof construction (see Figs 5.1 and 6.1 for more detail on bracing). For more

Goss's Roofing Ready Reckoner: From Timberwork to Tiles, Fifth Edition. C. N. Mindham.
© 2016 John Wiley & Sons, Ltd. Published 2016 by John Wiley & Sons, Ltd.

CE mark, ensures products meet the European Standards. This allows the product to be used in the European Economic Area.

BMTRADA is a UK based provider of assurance services to businesses around the world: including certification, testing, inspection and verification, technical training services covering a broad range of standards, schemes and products.

PEFC. Program for the Endorsement of Forest Certification. Promotes sustainable forest management through forest certification. This is an international standard which provides a chain of cusody certification.

UKAS. The United Kingdom Accreditation Service is the only Government recognised accreditation body in the UK, to measure certification bodies against international recognised standards, ensuring the certification body is fit for purpose.

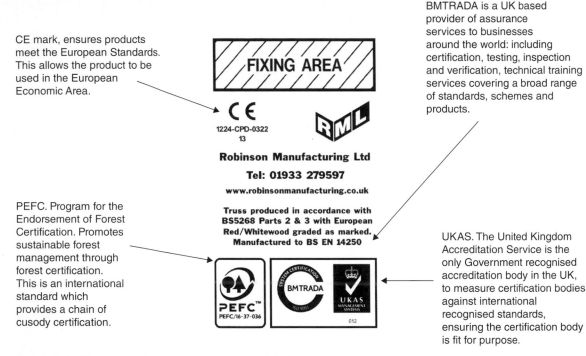

Figure 8.1 Trussed rafter quality standards label.

information on trussed rafters, from ordering through handling on site to erection on the building, contact the Trussed Rafter Association who, via their web site, can provide all of the information needed, including a list of approved manufacturers and a very helpful video. Contact information on the Association can be found in the Bibliography.

SPANDREL PANELS

A spandrel panel is a prefabricated triangular panel made to the general shape of the trussed rafter, and is placed at each end of the roof forming the inner skin of the external gable wall. The panel is of timber and sheet material construction, and has for many years been part of timber-framed house construction. The panels are now readily available as part of the trussed rafter whole-roof package. See book cover illustration for spandrel panels in construction.

For an external gable end, the panel would be clad with a suitable exterior quality building board such as Orientated Strand Board (OSB) and it would be faced with a water-resisting breather-type building paper to provide temporary waterproofing before the outer skin of brickwork can be built. The ties for the brickwork can be factory-fixed as flat stainless steel strips bent out and built into the brickwork as the courses of brick are laid. It must be remembered that the roof structure is supporting the brick gable end on a dwelling against wind pressure and suction forces. The brick gables DO NOT support the roof!

A typical spandrel gable end construction is shown in Figure 8.2, in which case the brick ties are omitted as they are the responsibility of the builder and not the roof package designer and manufacturer.

Spandrel panels can be used for the gable point of separating or party walls. In this case two panels would be used, one for each skin of the cavity wall below to provide fire resistance through the roof space of the adjacent buildings. The panels would be clad with two layers of 12.5 mm plasterboard with joints between the sheets staggered. If the party wall is for a dwelling room (as is the case with a room-in-the-roof), then additional fire proofing and sound proofing is required.

125

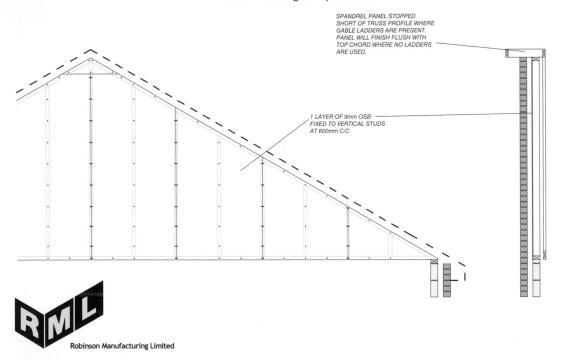

*SPANDREL PANEL STOPPED
SHORT OF TRUSS PROFILE WHERE
GABLE LADDERS ARE PRESENT.
PANEL WILL FINISH FLUSH WITH
TOP CHORD WHERE NO LADDERS
ARE USED.*

*1 LAYER OF 9mm OSB
FIXED TO VERTICAL STUDS
AT 600mm C/C*

Robinson Manufacturing Limited

Figure 8.2 Typical prefabricated spandrel panel.

CRANE-ON ROOF ASSEMBLIES – TRUSSED RAFTERS

Taking roof construction a stage further, it is practical to assemble the whole roof (without tiles or slates) on the ground and then to hoist the whole assembly onto the pre-fixed wall plates. (Wall plate fixing is shown in Fig. 5.1.) For timber-framed construction the wall plate is an integral part of the construction of the walls.

To use this method of building it is necessary to have enough space adjacent to the building shell, to set up a level replica of the wall plates on the building. This is easy if more than one identical house is being built, as the slab foundation of the next house can be used as roof assembly platform.

The advantages of the crane-on assembly are as follows:

- Safer working; there is no lifting of the trussed rafters up to wall plate.
- No need for scaffold at first floor level internally, or the need to lay the first floor.
- All bracing is easily installed.
- Gable ladders, fascia and barge boards can be fixed at ground level.
- Gable end panels (spandrels) can be fixed.
- Water storage tank and platform (if required can be installed).
- It avoids dependence on the availability of brick layers to build the gable end if prefabricated spandrels are used.
- Time is also saved by fitting the tile underlay and tile battens at ground level.

There is of course the cost of a crane to be considered, but in terms of lifting the roof assembly is a relatively light weight. A spreader beam (often available from the crane hire company) is needed and is usually positioned at the peak of the truss (ridge level) inside the building. The lifting chains pass through the underlay, which can be left loose for final fixing at the tiling stage. The trussed rafter manufacturer should be consulted if this method of building is to be used, as clearly the trussed rafters are being stressed differently to their normal load-carrying design.

CRANE-ON ROOFS – STRUCTURAL PANELS

Structural panel roofs consist of factory-made prefabricated rafters, top (ridge) and bottom (rail) to fit to wall plate, clad usually on their upper surface with a sheet material of plywood, OSB or some similar product. This is usually overlayed with the tiling underlay, fixed with temporary battens that thus affords a temporary weatherproof roof structure when in position on the building.

This construction is more suitable for room-in-the-roof dwellings, and often requires the use of purlins to take part of the roof load, especially if solid timbers are used, with their limitation on section size and length. The panels can be constructed using metal lattice timber flanged joists (which are now in common use for the floors of housing), used as rafters with special shaping at the ridge and eaves. The method of manufacture of the component is similar to that used to produce roof trusses, and indeed the product is available from many roof trussed rafter manufacturers.

Structural panels can be made on site provided that an engineer's design is prepared and adequate skill is available for assembly. This is clearly possible with stock timber being readily available from timber merchants. However, for larger spans prefabricated 'I' beams are available manufactured similarly to the lattice joist/rafter mentioned above, but with the lattice web replaced by a ply, OSB or other sheet material. These are produced as a standard stock product intended for use as floor joists, but they can be adapted on-site for ridge and eaves cuts, following the manufacturer's design information. A typical 'I' beam construction is shown in Figure 8.3.

CRANE-ON ROOF ASSEMBLIES – STRUCTURAL INSULATED PANELS

Structural insulated panels (SIPs) are all proprietary systems for producing a roof, and all are factory-made. In their simplest form they are a framed panel (as above) but with the addition of insulation (usually a waterproof foam board) between the rafters, and with a ceiling board on the underside. In this form it is often referred to as a 'cassette'. The cassette is best suited to room-in-the-roof dwellings, providing as it does instant weatherproofing and *almost* all the thermal insulation required by building regulations. The top-up insulation required to meet full Building

Double bevelled support plate

Restraint Strap (Cullen S or Simpson LSTA)

JJI-Joist/JJ-Beam ridge beam or support wall

18x200mm OSB splice block one side only, fix with 6 no. 3.35x65mm nails, clenched (optional)

JJI-Joist/JJ-Beam blocking panels (For ventilation guidance, see detail R10)

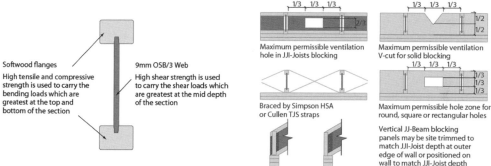

Softwood flanges

High tensile and compressive strength is used to carry the bending loads which are greatest at the top and bottom of the section

9mm OSB/3 Web

High shear strength is used to carry the shear loads which are greatest at the mid depth of the section

Maximum permissible ventilation hole in JJI-Joists blocking

Braced by Simpson HSA or Cullen TJS straps

Maximum permissible ventilation V-cut for solid blocking

Maximum permissible hole zone for round, square or rectangular holes

Vertical JJ-Beam blocking panels may be site trimmed to match JJI-Joist depth at outer edge of wall or positioned on wall to match JJI-Joist depth

Figure 8.3   Engineered timber composite 'I' beam and construction details.

Regulation compliance can be provided by fixing insulation-backed plasterboard beneath the cassette inner cladding. One company (Kingspan) produces SIPs using a rigid high thermal performance foam insulation bonded on both faces with OSB cladding, that can be used for both walls and roofs. Kingspan sells its SIPs through a system of nationwide delivery partners who can create the designs and provide whole-house packages for both self-build and large-scale developments. A typical Kingspan roof construction is shown in Figure 8.4.

Some cassette systems are capable of spanning from ridge to wall plate, but others require an intermediate supporting purlin, thus placing some load-bearing requirement on the gable end wall. A further alternative is to use structural panels of a lighter construction, spanning between the principal attic trusses at 1.2 m to 2.4 m centres. One such system using panels (cassettes) and attic trusses is known as the Structural Insulated Roofing System (SIRS), produced by Donaldson Timber Engineering. The cassettes give rigidity to the roof when fixed to their supporting attic trusses, and the gable ends are simply filled in with similar spandrel panels.

ATTIC TO ROOM-IN-THE-ROOF CONVERSION
To convert an attic or loft to living accommodation requires detailed design, planning and building regulation compliance, and is outside the scope of this book. Some suggestions for adapting the structure of the roof may be of assistance, however.

For a traditional cut roof where the rafters are supported on purlins, it may simply require the introduction of some additional strengthening of the floor, but we will return to that later. In the case of a trussed rafter roof, it is likely that some of the internal members (WEBS; see Fig. 2.2 for an illustration of a trussed rafter) will need to be removed to provide adequate living space, and in that case an engineer (possibly from the trussed rafter producer; see Fig. 8.1 and accompanying text) should be consulted. It is most probable that purlins will have to be introduced to support the rafters, and the supply of some additional support for the ceiling joists will also be necessary.

Engineered timber roofing components and prefabricated roofs

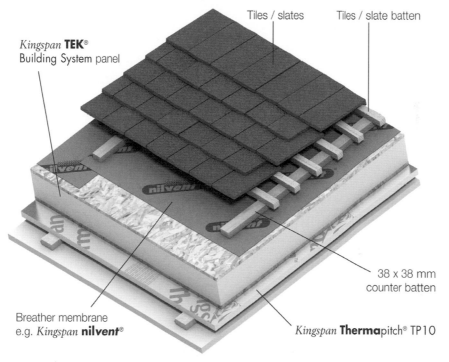

Tiles / slates

Tiles / slate batten

Kingspan **TEK**® Building System panel

38 x 38 mm counter batten

Breather membrane e.g. *Kingspan* **nilvent**®

Kingspan **Therma**pitch® TP10

Figure 8.4 Construction of a structural, insulated roof panel. With permission of Kingspan Insulation Ltd. Figure 4 from the Kingspan TEK Specification Manual (9th Issue, Feb 2014).

JES Installation

Step 1:

Use the JES as a template to mark the cut line and fastener hole positions on the end of the I-joist as shown, ensuring that the ends are flush.

Remove the JES before cutting drilling the I-joist.

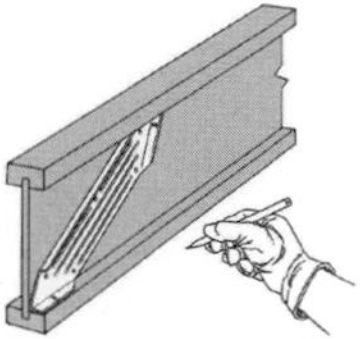

Step 2:

Cut and drill the I-joist. Use a 6 mm diameter drill bit to create the holes.

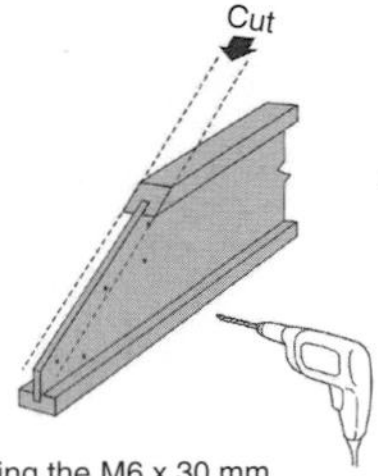

Step 3:

Securely install a JES on both sides of the I-joist using the M6 x 30 mm stainless steel Hex bolts and nylon washers supplied. Note the position of the lip on the JES which must be installed with the lip following the top-most edge of the adapted I-joist.

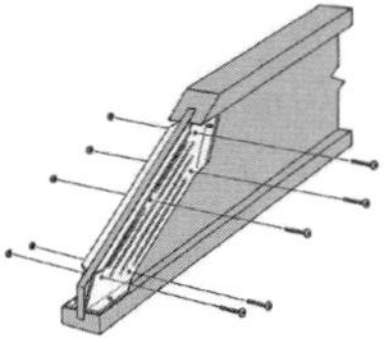

Step 4:

Position I-joist in between the existing trusses as shown, ensuring that a minimum 90 mm of end bearing is achieved.

Joist layouts will vary - please refer to engineer responsible for floor design.

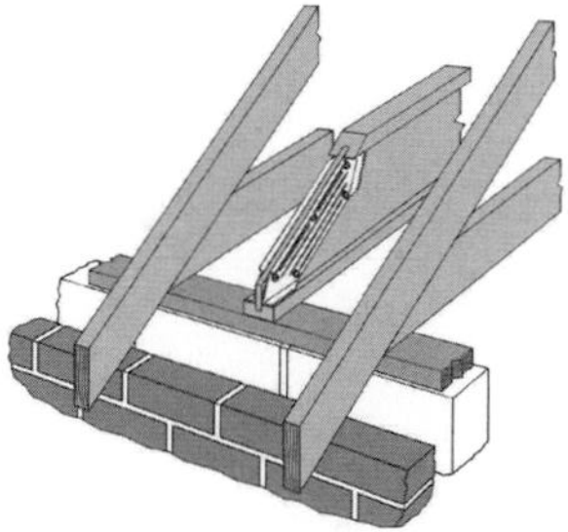

JES (Joist End Support)
Patent Pending

SIMPSON
Strong-Tie

Figure 8.5 Metal reinforcement fitting for 'I' beam used as attic floor joist.

Referring to the earlier sections of this chapter, the lattice joist produced by trussed rafter manufacturers, and also the 'I' joist, are both suitable for this situation. They are both lighter to handle than steel beams, and are both easier to pass services (plumbing and electrics) through the webs; in fact, the lattice joist needs no drilling at all. In addition, they can both be used as the additional floor joist mentioned above, and easily installed by the removal of the lower two courses of tiles, lifting the underlay, and sliding the man-handleable units into place. Some temporary support may be required to the ceiling joists as binders would have to be removed during the process. The ceiling joist can then be fixed to the new deeper joists.

The lattice joists would have to be specially made for the function as the eave ends would have to be built to suit the pitch of the roof. However, a timber 'I' beam can be cut to the required pitch on site and reinforced with metalwork specially designed for the purpose; this is produced by Simpson Strong-Tie, and is known as the 'Joist End Support' (JES). The installation of the JES is shown in Figure 8.5, and full fitting instructions are provided by the manufacturer.

9 ROOF COVERINGS – UNDERLAY, BATTENS AND TILES

UNDERLAY

Underlay is a thin sheet material that is laid over the top of the rafters and held in place by the roof covering supports or battens. There are principally two types of underlay: vapour-permeable and vapour-impermeable. The permeable type allows water vapour rising to the roof space to escape through the underlay, whilst not allowing any rain that may penetrate the roof covering to enter the building. The impermeable type is water- and vapour-proof in both directions, and can therefore give rise to moisture from water vapour being trapped within the roof space void. Ventilation of the roof in this case is essential (more information will be provided later).

Traditionally, roof underlays were a bitumen sheet reinforced with hessian-type cloth. This is still used but clearly falls into the impermeable type. The more modern underlays conform to the vapour-permeable standard. BS5534:2014 the Code of Practice for slating and tiling, came into force in February 2015, and makes specific demands of roof covering underlays, requiring them to be robust enough to resist wind suction uplift as well as being water-resistant. This has led to a new range of more complex underlays for tiling and slating. The new products are often multilayered products providing adequate strength to conform to the new requirements when fixed in accordance with the manufacturers' specifications. The fixing specification often changes with

Goss's Roofing Ready Reckoner: From Timberwork to Tiles, Fifth Edition. C. N. Mindham.
© 2016 John Wiley & Sons, Ltd. Published 2016 by John Wiley & Sons, Ltd.

the geographical location of the roof to be covered (see Chapter 3 for the effects of exposure on wind loading). The products will not only be waterproof but also vapour-permeable to allow the roof void to breathe and, in addition, some products now available also provide a degree of insulation. Permavent have a range of such products, 'Permafoil' being not only watertight, breathable and insulating but also providing fire resistance as extra protection – a very useful benefit when used under thatched roofs (for manufacturer's details, see Bibliography). The decision of which underlay to use will be influenced by the intended use of the roof void, whether storage or living loft, and if a 'cold' or 'warm' roof is being designed (see more information on this aspect under Roof Void Ventilation below).

Unless the roof structure is situated in Scotland, it is unlikely to be covered with a solid boarding or sarking. If it is, the roof underlay will be only partially supported by the sarking because counter-battening should be used to provide ventilation between the underlay and the sarking itself. For the majority of the UK, no sarking will be used and the underlay will be unsupported. Care must therefore be taken to avoid too much droop between the rafters, and particularly to avoid droop at the eaves behind the fascia. If this occurs, and rain penetrates the roof tiles, then ponding will occur behind the fascia, leading possibly to water penetrating adjacent to the wall plates and into the soffit area. An underlay support in some form of exterior boarding should be placed immediately behind the fascia to support the underlay. At any point on the roof, great care should be taken not to damage the underlay whilst battens and tiles are being laid on the roof (see Fig. 9.1 for underlay detailing).

The purpose of the underlay is to prevent any water that may penetrate the roof covering from entering the building, by draining it away into the gutter. It must therefore be applied by a method to ensure that the roof cannot be penetrated. As all underlays are produced in rolls, both vertical and horizontal joints in the material are unavoidable (except on roofs where the roll length exceeds the roof length between gables). Guidance on horizontal laps is provided in Figure 9.2.

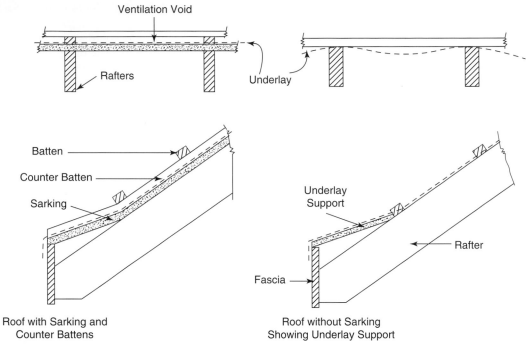

Figure 9.1 Underlay detailing.

Roof coverings – underlay, battens and tiles

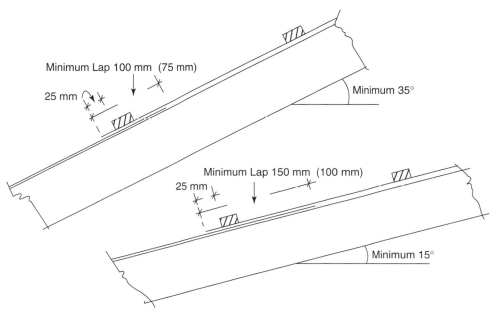

Figure 9.2 Minimum horizontal laps for unsupported and fully supported (shown in brackets) underlay.

This information is for not fully supported underlays. For further information on underlay support, see the illustrations in Chapter 10.

BATTENS

Battens support the roof covering of tiles, slates or shingles, spanning from rafter to rafter and provide a medium into which the tile fixing nails can be bedded. They also provide a 'ladder' for ease of walking on the roof, but care must be taken to stay along the rafter line and not to accidentally penetrate the underlay. Any damage to the underlay must be repaired in such a way as to avoid water penetration – sticking a piece on the surface will probably not be adequate.

The following aspects must be considered when specifying battens:

(a) The weight of the tiles or roof covering.
(b) The method of fixing required by the tiles, that is, a nib depth and nailing recommendation.
(c) The distance between the rafters, that is, the span of the battens,
(d) The durability of the material of the batten.
(e) The strength of the batten to conform to BS5534:2014.
(f) If in doubt, consult the tile or slate manufacturer, giving full details of the location of the building to be roofed.

In general, timber battens will be used as they are economical, readily available, easily handled, easy to nail to, and easy to cut. They are, however, generally of a small cross-sectional area, typically 38 mm wide × 25 mm deep or 50 mm wide × 25 mm deep. Therefore, they should be of good quality timber. Any defects, splits, knots or wane will greatly affect the strength of the batten and, as the roofer may end up standing on the batten between rafters, they must be considered as the rung of a ladder. The British Standard indeed requires that the battens are free from decay, insect attack, splits, shakes, knots or knot holes greater than one-third the width of the batten. Graded battens in accordance with BS5534:2014 are readily available from most timber and roofing merchants. They should be a minimum of 1200 mm in length and have both ends supported.

The size of the battens mentioned above should be checked with the technical literature of the roof covering tile or slate manufacturer, but in general terms 38 mm × 25 mm will span rafters of 450 mm centres, and 50 mm × 25 mm will span between 450 and 600 mm centres. If the roof covering is exceptionally heavy, such as stone slates as distinct from thin natural Welsh slate, then guidance should be sought from the provider or from an engineer.

Timber battens should be treated with preservative and are generally seen as requiring a minimum 60-year service life. The latest British Standard assigns a hazard class to building timbers, depending on their exposure to fungal and insect attack. Tiling battens are classified as hazard class 2. To place this in perspective, roof structural members are generally in hazard class 1, whilst members exposed directly to wetting, that is, the fascia, the barge and other exposed timbers such as exposed rafter feet, are in hazard class 3 and therefore need greater protection. By specifying the hazard class, the preservative treatment processing plant will be able to advise on the correct preservative system to be used. Buying BS5534:2014-compliant battens from a reputable merchant will ensure that the treatment specification is correct for purpose.

All preservatives contain chemicals to do their job, but some can prove corrosive to certain metal fixings. It is therefore essential to ensure that the nails used to fix battens to rafters, and the nails used to fix tiles to battens, are compatible with the preservatives used to treat the battens. Bright wire nails are NOT acceptable; galvanised nails may be, but check with the preservative supplier. Aluminium nails are commonly used, as are copper nails, especially for fixing tiles to battens. The tile manufacturer's recommendations should be sought to ensure nail (or screw) compatibility with the tiles or slates specified.

INSULATION AND VENTILATION

The drive to reduce carbon emissions, and therefore global warming, is causing the government to set increasingly higher standards of thermal insulation for all buildings, including dwellings. Part L of the *Building Regulations* sets out the requirements and has become a complex document, giving the standards to be achieved for the whole building (not just the roof or the walls, or any other element) for all new dwellings and refurbishment. However, different standards are set for domestic new and refurbishment structures. (See the last section in this chapter 'Building Regulations – Thermal Performance Requirements'.)

It is not intended in this text to describe the calculations as they involve the 'whole' building, and this book is concerned only with the roof structure. Guidance on the insulation of the roof structures is set out below. The largest area of heat loss in a building – assuming normal brick and block or brick and timber frame construction – is that through the roof, and it is for that reason that the standards indicate that the roof should offer the greatest resistance to heat loss. These standards are expressed as U values. To give an idea of the relative performance of the building element, the higher the value indicating the greater the heat transmission or loss, we have the following U values:

- External walls 0.18–0.24 W/m^2k
- Windows and doors 1.2–1.60 W/m^2k
- Floors 0.13–0.16 W/m^2k
- Roofs 0.13–0.16 W/m^2k

Part L of the *Building Regulations* deals with all aspects of the conservation of fuel and power. Whilst this concerns itself with the thermal performance of individual elements, it also recognises the effect on heat conservation caused by air changes within the building resulting from poor draught sealing; not only of the opening parts of windows and doors but also any gaps which may occur around the frames.

Paying great attention to the correct specification and installation of the insulation in the roof will have a major effect on the thermal efficiency, and therefore the cost of heating the building. This is particularly true of refurbishment and extension projects where the original insulation may be negligible.

Warning!

Do NOT specify the insulation without careful consideration to the possibility of condensation forming in the roof space – this can lead to wet insulation, and wet insulation DOES NOT offer any thermal resistance; it MUST be dry (see Roof Void Ventilation section below).

There are numerous types of insulation material available, all having their advantages and disadvantages in their ease of installation in a building and also, of course, their cost. Some are sold on their 'green' merits, being natural products and others are man-made, consuming energy to produce them. One product, for instance, made from sheep's wool uses only 14% of the energy in its manufacture compared to that taken to produce glass-fibre insulation. An indication of the insulation types available is set out below, but it must be stressed that this is by no means a complete list.

Rockwool, mineral wool, glass-fibre

A traditional and readily available product in roll or slab (bat) form. It is easily handled but some personal protection may be required in the form of gloves and masks.

Hi-tech multilayer reflective insulation

A high-performance for a relatively thin layer, easily applied in roll form, that reduces the risk of interstitial condensation (i.e., condensation forming within the insulation layer) because the surface layers are impermeable. (See Figure 9.3 for the use of Super FOIL, a multilayer insulation; see Bibliography for contact information.)

Polystyrene sheet

A lightweight rigid sheet particularly suitable for fixing between rafters in attic type constructions; check for fire-resistant standard.

Spray-applied polyurethane foam

A product usually applied by specialist contractors, but available as a DIY spray-on foam. This type is particularly suitable for increasing the insulation in refurbishment works where rafters are not of regular shape. This product also stops air leaks and air movements but of course is vapour-impermeable.

Rigid polyurethane sheet

Available in various thicknesses as a rigid sheet; it has applications similar to polystyrene sheet.

Blown-fibre products

Insulation is by inert fibres usually applied by specialist contractors, with the fibres being blown into position. This generally means that it is suitable for ceilings and not for sloping surfaces.

Fleece

Natural sheep's wool especially prepared for insulation. Available in rigid bat form.

Fibre insulation board

Rigid low-density fibre board, useful for attic applications between rafters. Available in various thicknesses.

SF60 SuperFOIL can be installed over or under the rafter and provides continuous insulation. It is ideal for roofs and attic conversions.

Over Rafter Application - Warm Roof

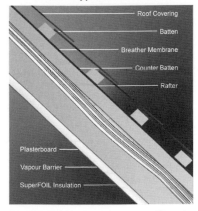

Roll out SuperFOIL over rafters, staple at least every 100mm and cover joins with SuperFOIL tape, overlap widths 75mm

At the eaves cut SuperFOIL around rafters and seal to cavity insulation or wall plate to create airtight envelope.

Fix battens parrallel with the rafters and apply breathable roof underlay according to manufacturers guidelines.

Fix roofing battens & tiles according to manufactures guidelines.

NB Use sarking board in Scotland

Under Rafter Application - Cold Roof

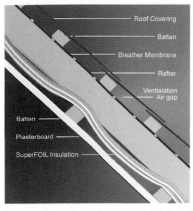

Roll out SuperFOIL, starting along the top of the roof. Batten horizontally over rafters. Staple at least every 100mm and cover joins with SuperFOIL tape, overlap widths 75mm

At the bottom of the roof pitch, staple the SuperFOIL directly onto the timber wall plate to create airtight envelope.

Fix battens across the rafters and ensure air gap between SuperFOIL and the plasterboard.

NB Use sarking board in Scotland

Figure 9.3   Specification of multilayer insulation in warm and cold roofs. SF60 Superfoil.

To give an indication of the different thicknesses of insulation material required to meet the basic building regulation requirements for roof structures, 270 mm of rockwool would be needed but only one layer of approximately 25 mm of hi-tech insulation plus 120 mm of mineral wool or 100 mm of polyurethane rigid sheet board. Whilst the rockwool insulation would probably be suitable for a roof ceiling insulation layer without loft space storage (there is no point in compressing the 270 mm by laying boards over the roof joists, as this defeats the insulation performance), the thinner materials (such as the polyurethane rigid sheet and multilayer insulation) will be more suitable for between-rafter insulation in an attic or loft structure. The cost as laid must also be taken into consideration, and whilst the basic cost of the insulation may be relatively economic, if there is a high wastage factor then this could affect the 'as-laid' cost of the insulation. For further information from various manufacturers of some of the above products, see the Bibliography.

Roof Void Ventilation

Why Ventilate?
Many old, pre-1900s houses had little or no insulation and, indeed, no underlay under the tiles or slates. Warm moist (vapour-laden) air rose through the building into the roof void and was vented out through the numerous small gaps between the tiles. The roof timbers where thus protected from moisture build up within the roof void (assuming that the roof covering was kept in good condition), and lasted for centuries. The gaps did of course let in insects, some of which attacked the timbers, but most wood worm (which is the most common form of attack throughout the majority of the country) came from infestation in cheap plywoods brought in during the early 1900s in furniture.

Ever-increasing standards to conserve energy result in higher levels of thermal insulation, often applied at ceiling level, which gives rise to a cold roof void. Vapour still rises from the house but now passes through the insulation and readily condenses into moisture on any cold surface within the roof space. This could simply be nails, metal connectors, and steel support systems that may be in the roof and, on trussed rafter roofs, the truss connector plates themselves. If there were still no underfelt within the roof space the majority of this moisture would be ventilated and would cause no problem. However, as there is often a bituminous felt underlay so common over the past several decades, this tends to keep the vapour within the building and can lead to the formation of so much condensation that it can drip from the cold surfaces onto the ceiling below. This can in extreme cases lead the occupier to think that the roof is leaking. The author has inspected roof spaces where this condition is so acute that frost has formed on the underside of the felt in the roof space, and when this thaws the amount of water descending on the ceiling can be significant. It is therefore necessary to ventilate this vapour from the roof void to keep the structure dry. If the project is a new construction, then building regulations set out rules for this ventilation, but it would be wise to follow these rules and provide ventilation in any refurbishment project even if that building is not subject to building regulations. This is simply good building practice.

Comparison of Warm Roofs and Cold Roofs

To better make provision for this ventilation, it must first be established whether the design is going to lead to a warm roof structure or a cold roof structure. A cold roof is that described above, that is, insulation at ceiling level giving a cold roof void. A warm roof void is that where the insulation is provided at rafter level, that is, as in an attic room, although it should be appreciated that there can be cold voids in an attic roof structure if there are spaces outside the habitable rooms. For instance, this could be the triangle which often occurs between the side wall and the eaves, and the small triangle which occurs above the ceiling up to the ridge. (See Figures 9.4 and 9.5 for warm and cold roofs.)

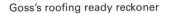
Goss's roofing ready reckoner

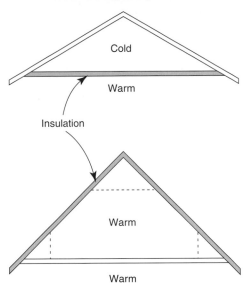

Cold

Warm

Insulation

Warm

Warm

Figure 9.4 Cold and warm roof voids.

Controlling Moisture-Laden Air Within the Structure

During the early 1900s houses described above, which had no underlay, the rafter level could be described as 'permeable'; that is, water vapour was free to escape through the gaps in the tiles. If, however, a bitumen-impregnated, cloth-based traditional roofing underlay is used, this is

Roof coverings – underlay, battens and tiles

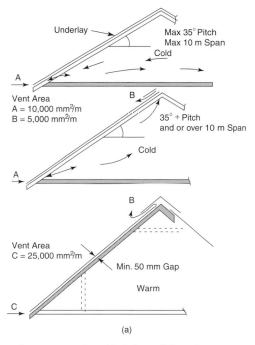

(a)

Figure 9.5 Designs for cold and warm roofs with (a) traditional non-permeable.

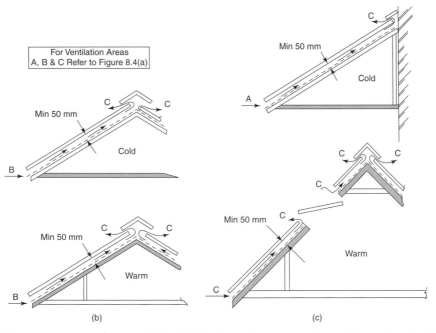

For Ventilation Areas
A, B & C Refer to Figure 8.4(a)

Figure 9.5 (b) vapour-permeable underlay. (c) Designs for mono pitch cold roof and attic warm roof with permeable underlay.

non-vapour-permeable and will not let vapour pass through. Today, however, a new generation of vapour-permeable underlays is available that will allow vapours to pass through but not allow water, which may penetrate the tiles into the building. There are standards set out in building regulations for the permeability of these underlays, and these should be checked against the product to be specified. In this case, the roof void may not need ventilating.

A vapour-permeable underlay works like ducks' feathers, in that it keeps out water from wind-driven rain which may pass through the tiles, but allows the 'body' of the house to breathe by exhausting vapour. If this concept is difficult to appreciate, then the reader should consider the effects of hard exercise in a waterproof non-breathable jacket: perspiration condenses on the inside of the material to make it wet inside, even when there is no wetting from the outside. This is exactly the same effect as the moisture within the house from bathing and cooking on the underside of an impermeable underlay. The new modern vapour-permeable underlays can be compared with the latest high-tech fabrics used on high-quality waterproof jackets, which allow the perspiration to breathe out without making the jacket wet inside, but do not let water in. Redlands 'Spirtech' 400 2S underlay is one such product (see Bibliography for contact details).

How do we Decide Which Roof Type to Specify?
Do we want a cold or a warm roof?
Do we need the roof void for storage?
> *Yes* – design a warm roof
> *No* – design a cold roof.
Do we need to live in the attic space?
> *Yes* – design a warm roof
> *No* – design a cold roof.
Various designs for cold and warm roof structures are illustrated in Figure 9.5.

In Scotland and Northern Ireland, where fully boarded roofs are common (i.e., boards on top of the rafters or 'sarking' is used), ALL underlays should be treated as impermeable and ventilation provided accordingly. This construction is rarely used elsewhere in the UK, but if it is the same rules apply as stated above. Refer to the *Building Regulations* for Scotland (Part C) and for Northern Ireland (Part B) for full details.

Economic Considerations

Heating Costs
Clearly, a roof insulated at ceiling level gives the least heated volume in the property, but full ventilation of the cold roof void and good insulation of the water storage tanks and all roof void water pipework is essential. If the roof void is to be used for a limited storage, one has to consider if the stored goods would suffer from the thermal extremes suffered within a roof void between the heat of the summer and the extreme cold of the winter. Also, even in extreme conditions with ventilation, stored metal objects may attract some temporary condensation in cold conditions.

A warm roof, however, results in a larger heated volume which has a proportional effect on cost, and perhaps with limited rafter depth it may prove more difficult to provide the required insulation. To gain this additional depth of rafter it may be necessary to apply battens to the underside of the rafter in the room area only to obtain a sufficient thickness of insulation. Alternatively, it will be necessary to use the higher-tech (and therefore likely more expensive) insulation material to fit between the rafters to gain the required insulation. In a warm roof, however, water storage tanks and piping do not need insulation (unless of course it is a hot water storage tank), but in such structures it is necessary to provide ventilation at the rafter level and at the ridge to ensure that moisture is not trapped within the rafters and insulation.

There follow some typical construction details for insulation and ventilation with vapour-permeable and vapour-impermeable underlays (Figs 9.6–9.8).

Roof coverings – underlay, battens and tiles

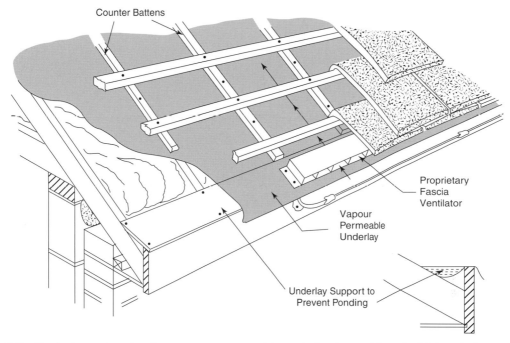

Counter Battens

Proprietary
Fascia
Ventilator

Vapour
Permeable
Underlay

Underlay Support to
Prevent Ponding

Figure 9.6 Typical construction for a vapour-permeable underlay in a cold roof with insulation at ceiling level (unventilated loft). Easy Verge Trim by Permavent.

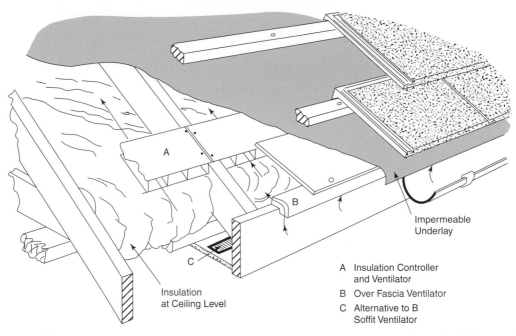

A Insulation Controller
   and Ventilator

B Over Fascia Ventilator

C Alternative to B
   Soffit Ventilator

Impermeable
Underlay

Insulation
at Ceiling Level

Figure 9.7 Typical construction for a vapour-impermeable underlay in a cold roof with insulation at ceiling level.

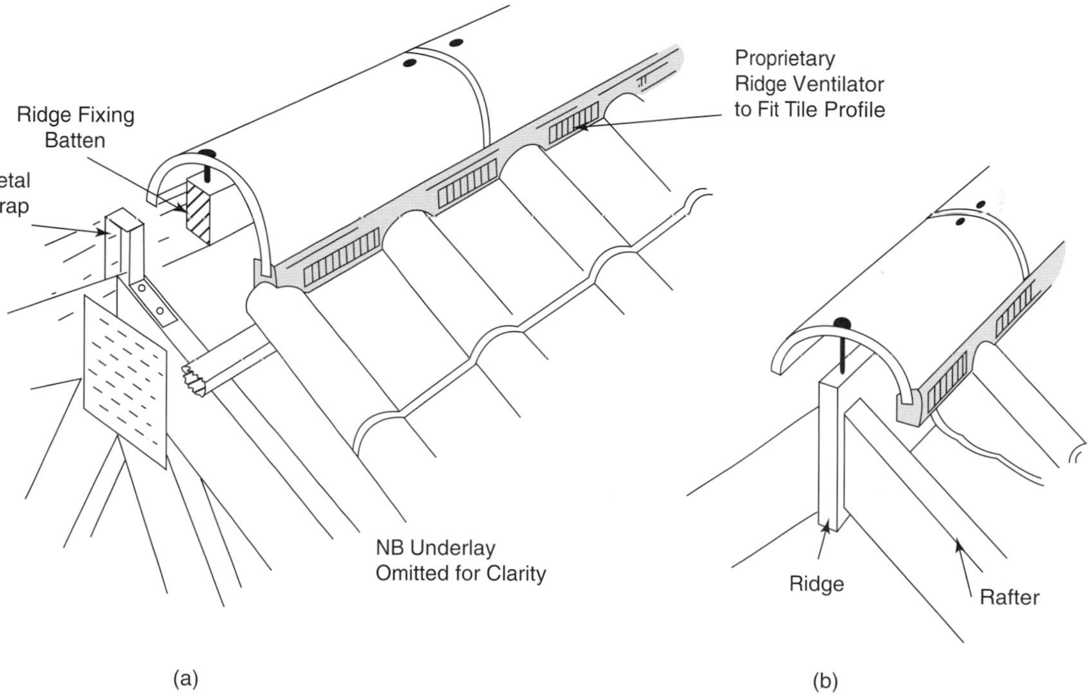

Ridge Fixing
Batten

Metal
Strap

Proprietary
Ridge Ventilator
to Fit Tile Profile

NB Underlay
Omitted for Clarity

(a)

Ridge

Rafter

(b)

Figure 9.8 (a) Typical ventilated ridge: trussed rafter roof. (b) Alternative ridge fixing: traditional cut roof.

CHOOSING THE ROOF COVERING

The choice of roof covering is affected by many factors, and consideration of each factor will influence the specification of the tiles, slates or shingles. The major items to be considered are as follows:

- The pitch or slope of the roof.
- The degree of exposure to wind and rain.
- The weight of the roof covering – what the roof structure is able to carry.
- Aesthetics.
- Local planning requirements, if any.
- The proposed life of the building.
- Building regulation requirements.
- Cost.
- Green issues.

Pitch or Slope of Roof

This may be determined by the architecture of the roof, the use which is proposed for the roof space (if any), or indeed the tile type to be used if this has already been specified by planning. Or, if on an extension, the need to match the existing roof covering.

The illustration in Figure 9.9 can provide a quick guide, and can be used to check the roof covering limitations for pitch; these limitations are for normal exposure. If the building is in an exposed coastal region or at high altitude, then the proposed tile or slate manufacturer should be consulted. The minimum pitch indicated for plain tiles and single lap interlocking tiles will generally increase as exposure increases.

Whilst slates may be satisfactory down to 20° pitch in moderate exposure as defined in BS5534:2014 conditions, they may be limited to a minimum of 40° pitch in severe exposed

Roof coverings – underlay, battens and tiles

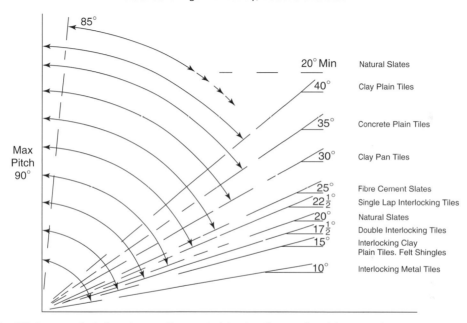

Figure 9.9   Minimum pitch for slates, tiles and shingles for roofs with normal exposure.

locations. However, Permavent have a slate roofing system which it claims will allow slates to be used down to a pitch of 12.5°, even in severe exposure conditions. This system is called the 'Low Pitch Slate System'. The company also has a roofing system for plain tiles, called 'Plain Easy' which allows plain tiles to be laid at a low pitch of 17.5°. The systems are complimented with matching products for valley, verge and abutments. (See Figure 10.4c for verge detail and Bibliography for contact information.) With plain tiles, slates and single interlocking concrete tiles, the adjustment can be made by increasing the lap of one tile over the other to improve the tile's performance. Double-interlocking tiles are those where there is an interlocking system on both the top of the tile and the side of the tile. These have definite maximum and minimum pitch levels, but because of the interlocking mechanism at the head of the tile they will generally be satisfactory at a lower pitch than will the single-side interlocking variety.

Degree of Exposure to Wind and Rain
This matter has been mentioned above, but is worthy of further consideration. What affects exposure?

(a) The degree to which the building is protected by other buildings and or trees.
(b) The height of the building above sea level.
(c) The proximity to the coastline.
(d) The location geographically in the UK.

If there is a choice of roof covering to be made, then obtain full technical information from a range of tile, slate or shingle manufacturers before starting to specify, and certainly before buying. As has been stated, the lap of the tile or slate will generally be increased with exposure, and this of course means that there will be an increase in number of tiles used on the roof, and therefore the associated cost. By choosing the double-interlocking tile, if possible (which may at first appear more expensive), the as-laid cost may in fact be less. The National House-Building

Council (NHBC) has guidance on this in their *Standards* document Chapter 7, Appendix 7.2-c, Table 1, which provides information on the minimum head lap. In the case of plain tiles, for instance, a 65 mm minimum lap is recommended for normal exposure, but this increases to 75 mm for severe exposure. Further information on the NHBC can be found in the Bibliography.

The effect of altitude (height above sea level) has been discussed in Chapter 3, under the heading 'How do we calculate the loading on the roof'. Proximity to the coastline is a geographical fact, and there are charts and maps available from the roof tiling manufacturers which indicate the wind exposure and indeed the rainfall exposure throughout the UK. All of this should be taken into consideration.

Loading on the Roof

Two situations must be considered. First, is the building a new building? Second, is the new roof covering being fixed to an existing roof?

On a new construction, clearly the roof timbers can be designed to accommodate the roof covering chosen and, if a cut roof, information in the tables of Chapter 3, or indeed the TRADA documents, can be used. However, for trussed rafter roofs the trussed rafter manufacturer will need to know the type of tile to be used in order to include this in the calculations for the roof structure. So, they will need to know if any solar panels, either photovoltaic (PV) or THERMAL are to be installed, and if so their proposed location on the roof (see Chapter 11 for information on solar panels).

Re-covering an existing roof structure will mean a structural check on the condition of the roof timbers and their fixings. Re-covering with the same type of tile or slate may be safe to do, but the minimum repair would be to replace all the existing battens and renew the underlay (if any was fitted originally), giving due consideration to the specification for the underlay with regard to potential condensation and insulation (this information was set out in the previous section).

Replacing original clay plain tiles with cheaper concrete plain tiles can save over 5% on the load on the roof, a fact which could be critical with an old roof structure. If there is any doubt about the safety of the roof structure, a structural engineer should be consulted and any suspect timbers replaced or supplementary timbers added to strengthen the structure.

One approach where other factors such as planning constraints do not demand like-for-like replacement, is to assess the safe load that the roof structure can carry by using the services of a structural engineer, and then select a covering within the roof's safe load-carrying capacity. A light roof covering such as asphalt (bitumen felt) or cedar timber shingles could be a consideration. The former are available in a range of colours and tile shapes to mimic traditional tiles. The weight of felt shingles which needs an under decking (or sarking) on which they are fixed, will only be approximately 20% of that of the weight of plain tiles. They do produce an impermeable layer, however, and therefore ventilation of the roof void is absolutely essential.

A quick guide to roof covering weights by type is shown in Figure 9.10. The data given here are an average; check with the manufacturer for exact loadings, and check the head lap required for the designed exposure, as this also increases the effective load of the tiles on the roof. This factor applies particularly to natural slates.

Aesthetics

How important are aesthetic considerations? Is the roof in an isolated area? What type of roof covering is characteristic of the locality? Is there a planning requirement imposed? Do you have to replace like-for-like on a renovation? Is there a colour preference? Do you wish to match a nearby existing roof? When all of the above applicable questions have been answered, the roof covering will have been specified or at least filtered down to a very limited range.

Local Planning Requirements

Always check with local planning authorities when replacing an existing roof covering, especially if the building is in a conservation area or indeed is itself a listed building. Most new buildings will require planning consent so, again, a check with the local planning officer will help make the

Roof coverings – underlay, battens and tiles

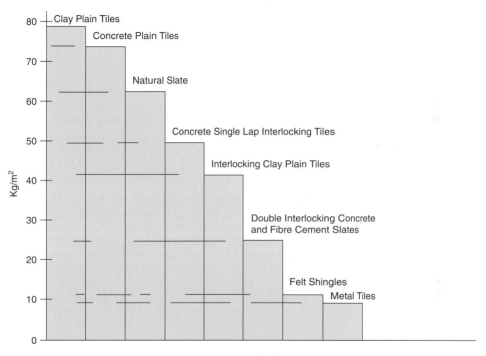

Figure 9.10 Approximate roof covering weights in kg/m².

choice of tile covering, if only by colour. It is too late – and can be very costly – if the individual's choice of re-roofing material conflicts with that of the planning guidelines, as the planners will have the authority to insist that the roof covering is replaced to their requirements.

Proposed Life of the Building

This may seem a strange consideration, but if you have the freedom of choice and the building's use is not long term, then some shorter-life roof coverings may be appropriate. Clay tiles for instance will last for 100 years, concrete tiles and slates for 50 years, fibre cement slates for 30 years, and bitumen felt sheeting for 10–15 years depending on their quality. All of the above data are of course affected by exposure to sun, wind and rain, and of course the level of maintenance provided.

Building Regulation Requirements – Material specification

The building regulations state the obvious, that the roof covering must:

- resist the passage of rain and snow to the inside of the building;
- not be damaged by rain or snow; and
- not transmit moisture due to rain or snow to another part of the building that may be damaged.

They also quote the appropriate British Standards codes of practice for precast concrete cladding (non-load-bearing) and for the design and installation of natural stone cladding and lining.

There is also a fire rating requirement where roof coverings pass from one occupancy to another, that is, semi-detached or terraced house situations, and/or where the roof is in close proximity to a boundary. If this situation exists, or the project under consideration creates any of the above conditions, consult the local authority building control department and obtain data on the proposed cladding to be used with regard to its performance. Concrete and clay tiles generally present no problems, but check for felt sheet and felt and timber shingles.

Cost

Three basic questions have to be asked before any comparison can be made between various roof coverings available.

- Is the project DIY; that is, the labour of laying and fixing the tiles is NOT to be part of the cost comparison?
- Is the project to employ professional roofers on a supply and fix basis?
- Is it the intention of the project to purchase the tiles direct, but employ a professional roofer to fix them?

In both situations where the building owner is intending to purchase the tiles direct, a careful check should be made on the minimum haulage charges, as these can be quite steep for relatively small quantities.

When considering the cost of the materials themselves (without the cost of fixing) there is surprisingly little difference in cost. Natural materials are generally more expensive than man-made ones when comparing on the square metre, as-laid basis. Clearly, there are more battens required for small plain tiles than larger interlocking concrete tiles and this too must be taken into account. If the project is DIY, then the speed of laying the tiles may be of consideration, with the larger double-interlocking concrete tiles being much faster to cover the roof than plain tiles. This aspect will also of course have an effect on subcontract fixing costs.

As a summary on cost, then, natural slates are likely to be the most expensive roof covering, moving down through man-made slates, to concrete plain tiles and then to double-interlocking tiles, which will be as little as 30% of the cost of the natural slate product. One then has to balance these costs against life expectancy. The final comparison will vary from project to project, because to carry out a true comparison the cost per square metre as-laid is only one factor. Hips, valleys, verges, ridges, cutting and fitting around dormers, roof windows, chimneys and so on, will also affect the cost of the project. Especially on DIY projects, do not overlook the cost of WASTE.

BUILDING REGULATIONS – THERMAL PERFORMANCE

Conforming to Building Regulations requirements is no longer a simple matter of stating 'U' values for the major elements of the building. There is now a need to provide whole-house performance for heat loss, which involves complex calculations. Quoted below is an overview of the current situation provided by Anthony Gwynne MRCS, MIFireE. author of *Guide to Building Control*. Details of this book can be found in the Bibliography.

Building Control

A separate system of building control applies in England, Wales, Scotland and Northern Ireland.

Building Control System in England

The Building Act 1984 and the Building Regulations 2010

The power to make building regulations are contained within Section 1 of the Building Act 1984, which deals with the powers of the Secretary of State to make building regulations for the following purposes:

• Securing the health, safety, welfare, and convenience of people in or about buildings.
• Conservation of fuel and power.
• Preventing waste, undue consumption, misuse or contamination of water.

The Building Act 1984 can be viewed at: www.legislation.gov.uk.

The current building regulations are the Building Regulations 2010 and The Building (Approved Inspectors, etc.) Regulations 2010 which came into force on October 1st 2010, and applies to England. A separate system of building control applies in Scotland, Northern Ireland and Wales. The 2010 Regulations in both cases consolidate the Building Regulations 2000 and the Building (Approved Inspectors, etc.) Regulations 2000, incorporating amendments since 2000. The Building Regulations are very short, contain no technical details, and are expressed as functional requirements and are difficult to interpret or understand. For this reason, the department for Communities and Local Government publishes guidance on meeting the requirements in a series of documents known as 'Approved Documents'.

Approved Documents

The Approved Documents are intended to provide guidance on how to achieve the requirements of the building regulations for dwellings and non-dwellings and make reference to other guidance and standards. In themselves, the Approved Documents are not mandatory and there is no obligation to adopt any particular solution contained within them if it can be achieved in some other way. In all cases it is the responsibility of the designer, applicant/owner and contractor to ensure the works are carried out in compliance with the building regulations. The current Approved Documents for dwellings are listed below and are available to view on the Department for Communities and Local Government web site: www.communities.gov.uk, or to purchase from The Stationary Office (TSO) on line at www.tsoshop.co.uk or telephone: 0870 600 5522. TRADA Technology span tables are available from: www.trada.co.uk/bookshop.

A: Structure (2004 edition with 2010 and 2013 amendments).
B: Volume 1: Fire safety in dwelling houses (2006 edition with 2010 & 2013 amendments).
C: Site preparation and resistance to contaminants and moisture (2004 edition with 2010 and 2013 amendments).

D: Toxic substances (1992 with 2002 and 2010 amendments).

E: Resistance to the passage of sound (2003 with 2004, 2010, 2013 & 2015 amendments).

F: Ventilation (2010 edition with further 2010 amendments).

G: Sanitation, hot water safety and water efficiency (2015 edition).

H: Drainage and waste disposal (2015 edition).

J: Combustion appliances and fuel storage systems (2010 edition with further amendments).

K: Protection from falling, collision and impact (including safety glazing which replaces Approved Document N) (2013 edition).

L1A: Conservation of fuel and power in new dwellings (2013 edition).

L1B: Conservation of fuel and power in existing dwellings (2010 edition incorporating 2010, 2011 & 2013 amendments). Domestic building service compliance guide (2013 edition).

M: Access to and use of buildings: Volume 1: Dwellings (2015 edition).

P: Electrical safety (2013 edition).

Q: Security – Dwellings (2015 edition).

Regulation 7: Materials and workmanship (2013 edition).

The building regulations – thermal performance requirements

Conforming to the thermal performance requirements of the Building Regulations for dwellings can be a complex matter, and unless you are experienced in construction you may need to get some professional advice/help from a suitably qualified and experienced property professional and energy consultant.

Thermal performance requirements for dwellings are contained within two Approved Documents, namely:

1. Approved Document L1A: Conservation of fuel and power in new dwellings (2013 edition).

2. Approved Document L1B: Conservation of fuel and power in existing dwellings (2010 edition with 2010, 2011 & 2013 amendments).

New dwellings (dwelling = a self-contained, single household)

Demonstrating compliance for the energy efficiency of new-built dwellings is a complex matter and takes into consideration the whole building. There is also a significant number of design considerations that need to be addressed by the building designer at the design stage.

Depending on the building type and specification used, additional renewable energy technologies may be required to achieve compliance; for example, PV panels. You should consult an accredited Energy Assessor for advice and always check with building control that your proposals comply with the requirements of the building regulations.

Conversions of buildings into dwellings – for example, the conversion of agricultural barns, is known as a 'Material Change of Use' and is contained in Approved Document L1B).

In ADL1A there are five separate criteria listed below that needs to be satisfied and checked using compliance software SAP 2012. The data are used to produce a report that should be sent to the building control body at the design stage to confirm that compliance has been achieved (see below for as built report required on completion).

Five criteria for establishing compliance as follows:

Criterion 1: Achieving the Target Emission Rate (TER) and Target Fabric Energy Efficiency Rate (TFEE).
Criterion 2: Limits on design flexibility.
Criterion 3: Limiting the effects of heat gains in summer.
Criterion 4: Building performance consistent with DER and DFEE rate.
Criterion 5: Provision for energy-efficient operation of the dwelling.

The above five criteria are considered in more detail as follows:

Criterion 1: Achieving the Target Emission Rate (TER) and Target Fabric Energy Efficiency Rate (TFEE)

At the design stage The Dwelling Emission Rate (DER) must be no worse than the corresponding Target Emission Rate (TER), and the Dwelling Fabric Energy Efficiency Rate (DFEE) must be no worse than the Target Fabric Energy Efficiency Rate (TFEE). On completion of the building, the final DER and DFEE must be based on the building as constructed. The TER/DER and TFEE/DFEE rate procedures/calculations to be carried out in accordance with paragraphs 2.8 to 2.30 of ADL1A.

The Target Emission Rate (TER) and Target Fabric Energy Efficiency Rate (TFEE)

The Target CO_2 Emission Rate (TER) and Target Fabric Energy Efficiency (TFEE) rate are the minimum energy performance requirements for a new dwelling approved by the Secretary of State in accordance with regulation 25.

The notional dwelling used to determine carbon dioxide and fabric energy efficiency targets is the same size and shape as the actual dwelling, constructed to a concurrent specification. The Part L 2013 specifications have been strengthened to deliver 6% carbon dioxide savings across the new homes build mix relative to Part L 2010.

The TER is expressed as the mass of CO_2 emitted in kilograms per square metre of floor area per year. The TFEE rate is expressed as the amount of energy demand in units of kilowatt-hours per square metre of floor area per year. The results are based on the provision and standardised use of specified fixed building services when assessed using approved calculation tools. The TER and TFEE rate for individual dwellings must be calculated using SAP 2012.

Note: *A summary of the Part L 2013 notional dwelling is published in Table 4 of ADL1A, with the full detail in SAP 2012 Appendix R. If the actual dwelling is constructed entirely to the notional*

166

dwelling specifications it will meet the CO_2 and fabric energy efficiency targets and the limiting values for individual fabric elements and buildings services. Developers are, however, free to vary the specification, provided the same overall level of CO_2 emissions and fabric energy efficiency performance is achieved or bettered.

The Dwelling Design Emission Rate (DER) and the Dwelling Fabric Energy Efficiency Rate (DFEE) The DER is the dwelling CO_2 emission rate expressed as $kgCO_2$ /(m^2.year) and the DFEE is the dwelling fabric energy efficiency rate expressed as kWh/(m^2.year) is calculated by an accredited Energy Assessor in accordance with the Governments Standard Assessment Procedure (SAP 2012)

Submission of information to building control body
- Before works commence: The design SAP report rating containing the TER/TFEE and DER/DFEE should be sent to the building control body at the application stage.
- On completion of the works: An as-built SAP report rating containing any changes together with an Energy Performance Certificate (EPC) should be sent to the building control body to fulfil building regulations requirements, and these can only be produced by an Accredited Energy Assessor.

Criterion 2: Limits on design flexibility
Limiting fabric standards (Worst acceptable fabric standards)
The thermal performance of building elements and the building services efficiencies must not fall below the minimum values as contained in paragraphs 2.31 to 2.37 of ADL1A. The worst acceptable fabric standards to be in compliance with Table 2 of ADL1A and the guidance table below. In general, to achieve the TER and TFEE rate a significantly better fabric performance than that set out in Table 2 is likely to be required.

Criterion 2 is intended to limit design flexibility, and to discourage excessive and inappropriate trade-offs. For example, individual building fabric elements with poor insulation standards being offset by renewable energy systems with uncertain service lives.

Guidance Table: Limiting fabric parameters (Worst acceptable fabric standards) (see Table 2 of ADL1A for full details).

Element	Worst acceptable fabric standard U-value-area weighted average (W/m²k)
Roof	U-value 0.20
Wall	U-value 0.30
Floor	U-value 0.25
Party wall	U-value 0.20
Swimming pool basin[1]	U-value 0.25
Windows, roof windows, glazed roof lights[2], curtain walling and pedestrian doors	U-value 2.00
Air permeability	100 m³/h·m² at 50 pa

Notes: 1 and 2. See Table 2 of ADL1A for full details.

Criterion 3: Limiting the effects of heat gains in summer

The dwelling should have appropriate passive control measures to limit the effect of heat gains on indoor temperatures in summer, irrespective of whether the dwelling has mechanical cooling in accordance with paragraphs 2.38 to 2.42 of ADL1A. Excessive solar gains should be checked in accordance with Appendix P of SAP 2012, and adequate levels of daylight are recommended to be maintained in accordance with BS 8206 -2 Code of Practice for Day Lighting

Note: The purpose is to limit solar gains and heat gains from circulation pipes to reasonable levels during the summer period, in order to reduce the need for, or the installed capacity of, air-conditioning systems. Criterion 3 should be satisfied even if the dwelling is air-conditioned.

Criterion 4: Building performance consistent with DER and DFEE rate

The performance of the dwelling, as built, should be consistent with the DER and DFEE rate in accordance with paragraphs 3.1 to 3.30 of ADL1A.

Criterion 5: Provision for energy-efficient operation of the dwelling

The necessary provisions for enabling energy-efficient operation of the dwelling should be put in place in accordance with paragraphs 4.1 to 4.3 of ADL1A.

Model designs

If the actual dwelling is constructed entirely to this model specification it will meet the TER and, better, the TFEE rate and therefore pass Criterion 1. Table 4 of ADL1A and the guidance table below provide a summary of the concurrent notional building specification. More detailed information can be found in SAP 2012 Appendix R.

It should be noted, however, that the concurrent notional building specifications are not pre-scriptive and may not be the most economic specification in every case. Designers are free to explore the most economic specification to meeting the TER and TFEE rate in each case, provided that this specification meets all other provisions within this approved document, in particular the limiting fabric parameters in Table 2 of ADL1A.

Some builders may prefer to adopt model design packages rather than to engage in design for themselves. Such model packages of fabric U-values, boiler seasonal efficiencies, window opening allowances and so on should, if suitably robust, help the builder achieve compliance. The construction industry may develop model designs for this purpose and make them available on the Internet at: www.modeldesigns.info.

It will still be necessary to demonstrate compliance in the particular case by going through the procedures described in paragraphs 2.8 to 2.17 of ADL1A.

Guidance Table: Summary of the concurrent notional building specification (see Table 4 of ADL1A for full details).

Element or system	Values
Opening areas (windows and doors)	Same as actual dwelling up to a maximum proportion of 25% of total floor area[1]
External walls (including opaque elements of curtain walls)[6]	0.18 W/(m²k)
Party walls	0.0 W/(m²k)
Floor	0.13 W/(m²k)
Roof	0.13 W/(m²k)
Windows, roof windows, glazed roof-lights and glazed doors	1.4 W/(m²k) (whole window U-value)[2] g-value = 0.63[3]
Opaque doors	1.0 W/(m²k)
Semi-glazed doors	1.2 W/(m²k)
Airtightness	5.0 m³/(h·m²)
Linear thermal transmittance	Standardised psi values – see SAP 2012 Appendix R, except use of y = 0.05 W/m²k if the default value of y = 0.15 W/m²k is used in the actual dwelling
Ventilation type	Natural (with extraction fans)[4]
Air-conditioning	None

Guidance Table: (*Continued*)

Element or system	Values
Heating system	Mains gas if combi boiler in actual dwelling, combi boiler; otherwise regular boiler:
	Radiators
	Room sealed
	Fan flue
	SEDBUK 2009 89.5% efficient
Controls	Time and temperature zone control[5]
	Weather compensation
	Modulating boiler with interlock
Hot water storage system	Heated by boiler (regular or combi as above)
	If cylinder specified in actual dwelling, volume of cylinder in actual dwelling
	If combi boiler, no cylinder. Otherwise 150 litres
	Located in heated space
	Thermostat controlled
	Separate time control for space and water heating
Primary pipework	Fully insulated
Hot water cylinder loss factor (if specified)	Declared loss factor equal or better than $0.85 \times (0.2 + 0.051\ V2/3)$ kWh/day
Secondary space heating	None

(*continued*)

Guidance Table: (*Continued*)

Element or system	Values
Low-energy lighting 100% low-energy lighting	100% low-energy lighting
Thermal mass parameter	Medium (TMP = 250)

Notes: 1. The Building Regulations do not specify minimum daylight requirements. However, reducing window area produces conflicting impacts on the predicted CO_2 emissions: reduced solar gain but increased use of electric lighting. As a general guide, if the area of glazing is much less than 20% of the total floor area (TFA), some parts of the dwelling may experience poor levels of daylight, resulting in increased use of electric lighting.

2. The orientation of the elemental building is the same as the actual building. In plotting buildings onto a site the designers should consider the benefits of orientating buildings to the south (with large windows orientated south and smaller windows orientated north) to benefit from passive solar gains through having lower space heating demands. Designers should be aware of the risk of overheating through excessive solar gain in the summer and design shading to avoid excessive summer heat gain.

3. Higher g-values would also comply with the recipe as increasing solar gains reduces the space heat load. However, designers should be aware of the impact of g-value on the risk of overheating and optimise their choice accordingly. The U-value is set to 1.5 W/(m²k) for curtain walling glazed areas, as an allowance for thermal bridging.

4. See SAP 2012 Section 11: two fans for TFA up to 70 m²; three fans for 70 < TFA <100 m²; four fans for TFA > 100 m². A recipe approach can be followed if extractor fans are replaced with the same number of passive vents.

5. In order for a system to be specified with time and temperature zone control, it must be possible to programme the heating times of at least two heating zones independently, as well as having independent temperature controls. These two heating zones must be space heating zones. For single-storey open-plan dwellings in which the living area is greater than 70% of TFA, sub-zoning of temperature control is not appropriate and the recipe will default to programmer and room thermostat.

Guidance Table: U-values for party walls (see Table 3 of ADL1A for full details).

Party wall construction	U-value
Solid walls	0.0 W/m²k
Unfilled cavity with no effective edge sealing	0.5 W/m²k
Unfilled cavity with effective edge sealing around all exposed edges and in line with insulation layers in abutting elements	0.2 W/m²k
A fully filled cavity with effective edge sealing around all exposed edges and in line with insulation layers in abutting elements	0.0 W/m²k

Note: An unfilled cavity with no effective edge sealing (U-value of 0.5 W/m²k) will not achieve the minimum required worst case U-value of 0.2 W/m²k and will cause the dwelling to fail.

Typical insulation details for pitched roofs
Follow the construction details in Part A of Section 2 of this guidance for Domestic Extensions.

Roof insulation and ventilation gaps
Insulation to be fixed in accordance with manufacturer's details and must be continuous with the wall insulation but stopped back at eaves or at junctions with rafters to allow for a continuous 50 mm air gap above the insulation to underside of the roofing felt where a non-breathable roofing felt is used or 25 mm air space to allow for sag in felt if using a breathable roofing membrane in accordance with the manufacturer's details.

Guidance Table: Insulation fixed between/under rafters (typically tiles, battens, breather membrane, rafters at 600 mm centres, insulation layer fixed between rafters, vapour check and integral/12.5 mm plasterboard fixed under rafters).

Insulation product U-value 0.13 W/m²k	k-value	Required thickness of insulation (mm) and position in roof*
Celotex FR5000 and	0.021	120 mm FR5000 between rafters and 50 mm
Celotex PL4000	0.022	PL4000 insulated plasterboard
Kingspan Kooltherm K7	0.020	120 mm K7 between rafters and 62.5 mm
Pitched Roof Board and	0.020	K18 insulated plasterboard
Kingspan Kooltherm K18 insulated plasterboard		

*All unvented roofs using vapour-permeable underlay.
Notes: 1. Insulation to be installed in accordance with manufacturer's details. Source: Manufacturer's details.

Guidance Table: Insulation laid horizontally between and over ceiling joists (typically tiles, battens, breather membrane, rafters and ceiling joists at 600 mm centres, ventilated roof void, insulation layer fixed between/over ceiling joists and vapour-checked plasterboard fixed under rafters).

Insulation product U-value 0.13 W/m²k	k-value	Required thickness of insulation (mm) and position in roof*
Rockwool Roll or Earthwool	0.044	150 mm between joists and 200 mm laid over joists

*All unvented roofs using vapour-permeable underlay.
Note: 1. Insulation to be installed in accordance with manufacturer's details. Source: Manufacturer's details.

Guidance Table: Insulation fixed between/over rafters (warm roof construction; typically tiles, battens/counterbattens, breather membrane, insulation layer fixed over rafters, rafters at 600 mm centres, insulation layer fixed between rafters, cavity and vapour-checked plasterboard fixed under rafters).

Insulation product U-value 0.13 W/m²k	k-value	Required thickness of insulation (mm) and position in roof*
Celotex FR5000	0.021	80 mm fixed over rafters and 80 mm fixed between rafters, with 12.5 mm plasterboard fixed to underside of rafters
Kingspan Kooltherm K7 Pitched Roof Board	0.020	80 mm fixed over rafters and 80 mm fixed between rafters, with 12.5 mm plasterboard fixed to underside of rafters

*All unvented roofs using vapour-permeable underlay.
Note: 1. Insulation to be installed in accordance with manufacturer's details. Source: Manufacturer's details.

Pitched roof ventilation requirements when using a non-breathable roof membrane

(i) *Duo pitched roof with horizontal ceilings and insulation at ceiling level*
Roof insulation to be continuous with the wall insulation but stopped back at eaves or at junctions with rafters to allow a 50 mm minimum air gap. Cross-ventilation is to be provided by either proprietary fascia ventilation strips or soffit vents to opposing sides of roof at eaves level and fitted with an insect grill with a ventilation area equivalent to a 25 mm continuous gap for roof pitches below 15° or a 10 mm gap for roof pitches above 15°.

When the roof span is more than 10 metres or when the pitch is more than 35°, provide additional high-level ventilated openings equivalent to a continuous 5 mm air gap at ridge level to cross-ventilate roofs using proprietary dry ridge systems or vent tiles spaced and fixed in accordance with the tile manufacturer's details.

(ii) *Mono pitched roofs with horizontal ceilings and insulation at ceiling level*
Roof insulation to be continuous with the wall insulation but stopped back at eaves or at junctions with rafters to allow a 50 mm minimum air gap. Cross-ventilation is to be provided by either proprietary fascia ventilation strips or soffit vents at eaves level and fitted with an insect grill with a ventilation area equivalent to a 25 mm continuous gap for roof pitches below 15° or a 10 mm gap for roof pitches above 15°.

Provide high-level ventilated openings fitted with an insect grill equivalent to a continuous 5 mm air gap to cross-ventilate roofs using proprietary ventilation systems or vent tiles spaced and fixed in accordance with the tile manufacturer's details.

(iii) *Duo pitched roof with insulation following slope of rafters (rooms in the roof)*
Roof insulation to be continuous with the wall insulation but stopped back at eaves or at junctions with rafters to allow a continuous 50 mm air gap between the top of the insulation and underside of the roof membrane. Cross-ventilation to be provided by proprietary eaves ventilation strips equivalent to a 25 mm continuous air gap to opposing sides of roof at eaves level, fitted with insect grill and at ridge/high level to provide ventilation equivalent to a 5 mm air gap in the form of proprietary dry ridge system or vent tiles spaced and fixed in accordance with the tile manufacturer's details.

Proprietary vapour-permeable roof membrane

Ventilation to the roof space may be omitted, only if a proprietary British Board of Agreement (BBA or other third party accredited) vapour-permeable breathable roof membrane is used.

Vapour-permeable breathable roof membranes must always be installed in strict accordance with the manufacturer's details. Note: Some breathable membranes may also require additional roof ventilation in accordance with manufacturer's details.

FLAT ROOF CONSTRUCTION

Guidance Table: Insulation fixed between/under flat roof joists (vented 'cold roof'; typically waterproofing layer, timber deck, 50 mm ventilated cavity, firring strips, flat roof joists at 600 mm centres, insulation layer fixed between joists, vapour check and insulated plasterboard fixed under joists).

Insulation product U-value 0.13 W/m²k	k-value	Required thickness of insulation (mm) and position in roof
Celotex FR5000 and Celotex PL4000	0.021 0.022	150 mm FR5000 between joists and 50 mm PL4000 Insulated Plasterboard
Kingspan Kooltherm K7 Pitched Roof Board and Kingspan Kooltherm K18 Insulated Plasterboard	0.020 0.021	150 mm K7 between rafters and 52.5 mm K18 insulated plasterboard

Notes: 1. Insulation to be installed in accordance with manufacturer's details.

2. The joist depth must be sufficient to maintain a 50 mm air gap above the insulation and cross-ventilation to be provided on opposing sides by a proprietary ventilation strip equivalent to a 25 mm continuous gap at eaves level with insect grill for ventilation of the roof space.

Guidance Table: Insulation fixed above flat roof joists (Non-vented 'warm roof'; typically waterproofing layer, insulation layer, vapour-control layer, timber deck, cavity between joists, flat roof joists at 600 mm centres, and vapour-checked plasterboard fixed under joists).

Insulation product U-value 0.13 W/m²k	k-value	Required thickness of insulation (mm) and position in roof
Celotex EL3000 (or TC3000)	0.025 to 0.027	130 mm + 50 mm insulation in two layers (total 180 mm), boards mechanically fixed to the top of the flat roof joists using thermally broken fixings – roof coverings can be hot-bonded or torched on as appropriate. 12.5 mm plasterboard fixed to underside of joists
Kingspan Thermaroof TR27 LPC/FM	0.024 to 0.026	130 mm + 40 mm bonded together in two layers (total 170 mm thick)

Note: 1. Insulation to be installed in accordance with manufacturer's details.

From the above detailed overview it can be seen that it may well be advantageous to employ a specialist in this field, to prepare thermal performance data for any building project you may have in mind which requires Building Regulation approval.

10 ROOF COVERINGS – BUILDING DETAIL DRAWINGS

THE NEW BS 5534:2014

BS 5534 is the code of practice for slating and tiling of pitched roofs and vertical cladding. The code was first published in 1942 as CP (Code of Practice) 142. The latest version BS 5534 came into force in April 2015. The new requirement for tiling battens has been dealt with in Chapter 8, and here we look at the codes requirement for keeping the tiles or slates on the roof.

There is now a requirement to mechanically fix the ridge, the valley, the hip and a much more rigorous requirement to fix the individual tiles or slates. Damage to the perimeter of a roof by high winds either due to pressure or equally suction can lead to the failure of one inadequate fixed ridge, verge or roof tile, which quickly spreads once the wind is into the structure, leading to failure of the underlay (see Chapter 9) and to progressive failure of the roof coverings.

The precise specification for fixing the roof coverings will vary depending on the type of tile or slate, be they of clay or cement composite (possibly timber shingles) and also be building location-specific. This reflects the varying exposure to wind from low, at low levels and more central locations, to high, for higher altitudes and coastal regions. By way of illustrating this variation in fixing requirements, we will look at two situations – a two-storey house inland in middle England, and a similar house to the west of the country. On the Midland house all ridges, valleys and

Goss's Roofing Ready Reckoner: From Timberwork to Tiles, Fifth Edition. C. N. Mindham.
© 2016 John Wiley & Sons, Ltd. Published 2016 by John Wiley & Sons, Ltd.

hips must be mechanically fixed. All perimeter tiles – that is, the ridge, verge and eaves, as well as those around dormers and chimneys – must be twice-nailed with ALL other tiles once-nailed, with 50 mm × 3.5 mm aluminium alloy clout nails (or stainless steel). The location of the nail in the tile, whether the right hand or left hand, will depend on the tile type (refer to the manufacturer's specification for this information).

For the same house towards the West coast, the requirement for ridge valleys and hips are the same, but for the tiles, reflecting the higher wind exposure that can be expected, perimeter tiles must be nailed and clipped to the structure. Additionally, a band of full tiles adjacent to the perimeters must also be clipped in an alternative diagonal pattern. All other tiles must be nailed using 70 mm × 3.75 mm nails as specified above.

MANUFACTURER'S SPECIFICATION

The above aims to serve only as an illustration of the degree of fixing now required and the degree of variation geographically within the UK. All tile manufacturers will provide fixing specifications for their tile in the proposed building location, tailored to the individual job. Redland, for instance, have produced an 'on-line' service entitled 'Fix Master' which can be accessed by mobile phone. Clearly, this information must be to hand together with the correct clips, special flexible adhesive underlays for hips, valley, ridges and abutments and, of course, the nails and screws before tiling commences.

HIPS, RIDGES, VERGES AND VALLEYS – TRADITIONAL AND NEW METHODS OF CONSTRUCTION

Traditional methods for the above roofing elements have been typically illustrated in Figures 10.1, 10.2 and 10.3. The modern methods dictated by the need for better performance (BS 5534:2014), and also by the desire for greater efficiency on site, have led to a very wide range of dry fixing (without the need for conventional cement and sand mortar) methods, often specific to the tile

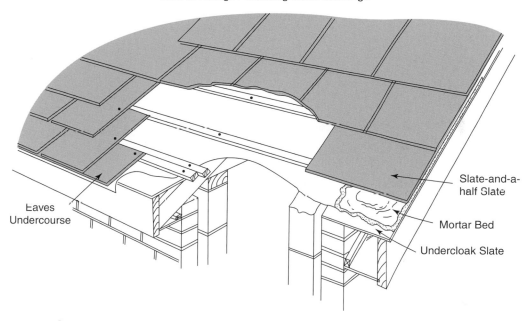

Slate-and-a-half Slate

Mortar Bed

Undercloak Slate

Eaves Undercourse

Figure 10.1a Slate roof. Eaves and verge.

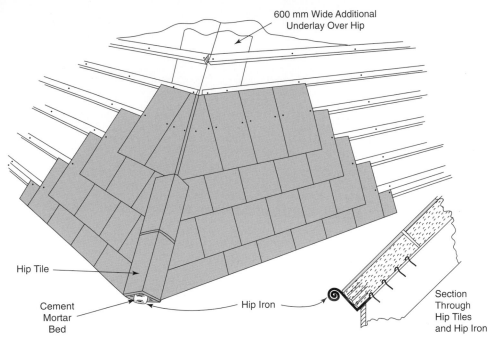

Goss's roofing ready reckoner

600 mm Wide Additional
Underlay Over Hip

Hip Tile

Cement
Mortar
Bed

Hip Iron

Section
Through
Hip Tiles
and Hip Iron

Figure 10.1b Slate roof. Hip detail.

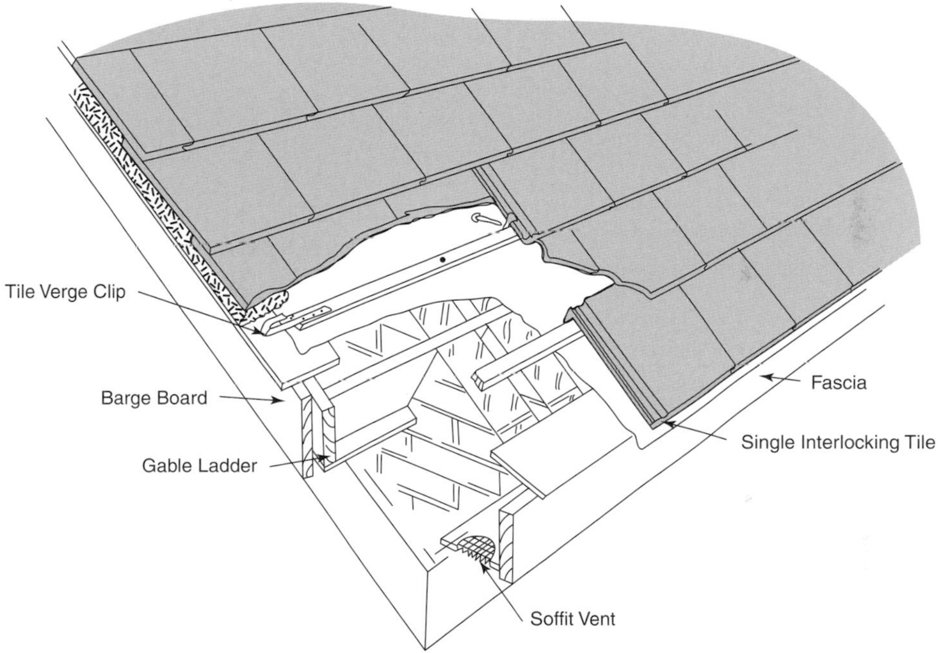

Roof coverings – building detail drawings

Tile Verge Clip

Barge Board

Gable Ladder

Fascia

Single Interlocking Tile

Soffit Vent

Figure 10.2a Single interlocking tile. Barge verge and eaves detail.

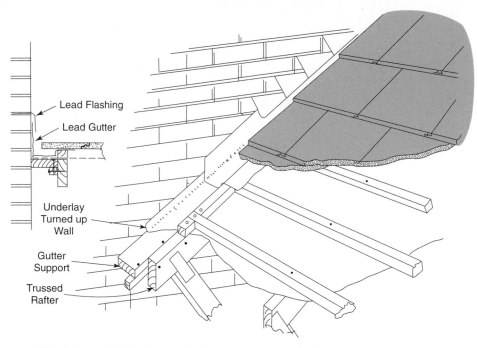

Lead Flashing

Lead Gutter

Underlay
Turned up
Wall

Gutter
Support

Trussed
Rafter

Figure 10.2b Single interlocking tile. Side abutment gutter detail.

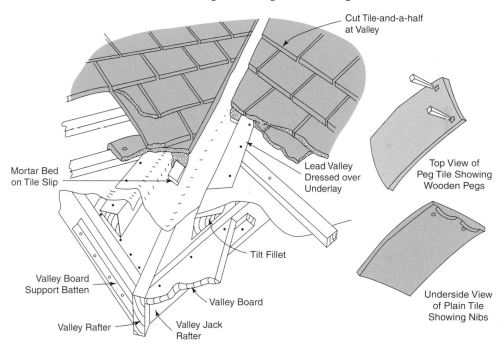

Cut Tile-and-a-half
at Valley

Mortar Bed
on Tile Slip

Lead Valley
Dressed over
Underlay

Top View of
Peg Tile Showing
Wooden Pegs

Tilt Fillet

Valley Board
Support Batten

Valley Board

Valley Rafter

Valley Jack
Rafter

Underside View
of Plain Tile
Showing Nibs

Figure 10.3a   Plain tile. Valley detail.

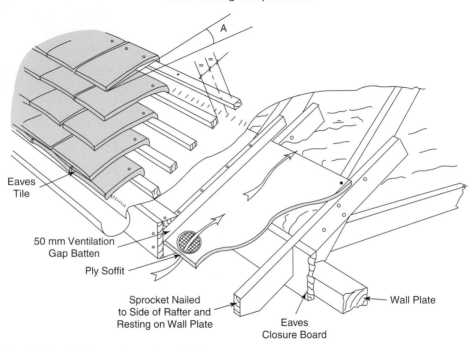

Goss's roofing ready reckoner

A

Eaves Tile

50 mm Ventilation Gap Batten

Ply Soffit

Sprocket Nailed to Side of Rafter and Resting on Wall Plate

Eaves Closure Board

Wall Plate

Figure 10.3b Plain tile. Sprocket eaves detail.

type and certainly specific to tile manufacturer. In general, they will have a plastic type undertray particularly at valley, abutments and eaves, often with some metal straps or clips, and they will all be screwed or nailed to the structure. The plastic-moulded undertrays also replace the need for lead in many cases. More on this subject later.

Figure 9.8 gives a generic illustration of a dry fixed ridge, while Figures 10.4(a, b and c) provide a sample of the solutions available from some of the roof covering manufacturers. Whilst tile and slate manufacturers all have systems to suit their specific products, companies such as Permavent product an extensive range of dry fixing plastic extrusion systems which are more universal, for a wide range of tile and slate types (see Bibliography for product details).

ROOFING PUTTIES

Traditional cement mortar is liable to shrink and over time, and lose adhesion to the items to which it is bedded, leading to loose tiles, ridges and hips. Now, with mechanical fixing being required dry fixing has become common place. However, if combined with roof putties a traditional appearance can be maintained. Roofing putties are also very useful for repairs, their main advantages being:

- No need for sand, cement, or water.
- Instant adhesion (no setting time).
- Instantly waterproof (although they cannot be bonded to a wet roof).
- Available in a range of colours to blend with the tiles or existing mortar.
- Conveniently packed to avoid waste.
- Good shelf life while sealed in packs.
- Little or no waste.

The method of application is to ensure that the surfaces to be bonded are dry and brushed free of dust. The putty is applied to the tiles, one ridge or verge tile at a time, and bedded down the ridge to the required location. Any excess putty can be knifed off, remoulded into a strip and

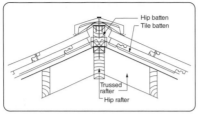

1a | Standard Rapid Hip

Underlay and batten the roof with ends of tiling battens supported on the rafter.

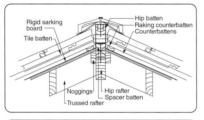

1b | Rapid Hip with Rigid Sarking

Finish sarking board at the side of the hip rafter. Fix a spacer batten to the hip rafter to a height level with the top of the counterbattens. Underlay, counterbatten and batten the roof with the tiling battens supported on the spacer batten.

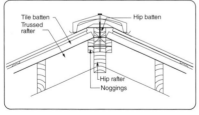

1c | Cambrian Rapid Hip

Fix 50 × 25mm noggings to the sides of the hip rafter. Underlay the roof. Fix a hip batten to the hip rafter at 300mm centres with 100 × 3.75mm ring shank nails provided. Batten the roof with ends supported on noggings. Go to step 5.

Figure 10.4a Redland Rapid HIP System.

Universal RidgeFast dry ridge system

- maximum rafter pitch 60°
- provides 5,000mm^2/m free vent area at ridge apex
- ensure gap is provided in roof underlay to vent roof void
- use one or two thicknesses of 50mm × 25 mm batten to fit batten brackets
- mechanically fix all top course tiles
- use block end ridge tile at ridge end

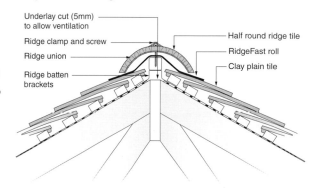

Underlay cut (5mm) to allow ventilation
Ridge clamp and screw
Ridge union
Ridge batten brackets
Half round ridge tile
RidgeFast roll
Clay plain tile

Figure 10.4b Marley Eternit Universal Ridge Fast Dry Ridge System.

reused on the next tile. The joint is then tooled off to give a smooth finish. Note that the tiles, ridges and hips must still be mechanically fixed to roof – the putty is NOT a substitute for screws or nails. One such product is 'Flexim' roofing putty, and another is 'Rapid Roof Putty' by Redland (see Bibliography for contacts).

NATURAL SLATES
Natural slate is one of the oldest roof covering materials, and although much is now imported slate quarries still exist in Wales and Cornwall, producing traditional riven slates of variable sizes and colours. Slate is an extremely durable material, lasting up to 100 years, and whilst 'slate grey' is a term used to describe the traditional slate colour they are also available in a range of colours,

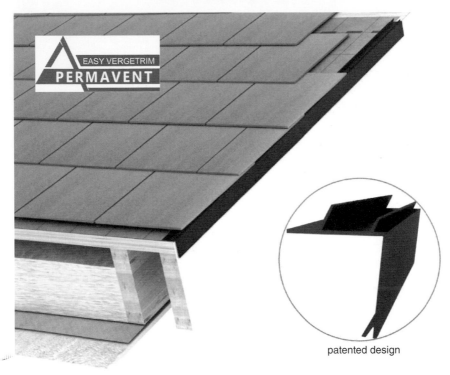

patented design

Figure 10.4c   Permavent Easy Verge Trim.

from the traditional grey through blue and green to heather shades. Laid on battens in the normal way, because the slates have no nibs to hold the slate onto the batten, they must be securely fixed with aluminium or copper silicone bronze nails. The holes for the nails are made by the roof layer using a slate holing machine; the position of the holes from top to bottom or 'head to tail' in slating terminology, is decided by the gauge to which they are to be laid. This gauge will depend on the roof pitch and exposure. The more severe the exposure, the more the slates will be overlapped one on the other, and will at some point on the course be three slates thick. Guidance on the lap and gauge of the slates can be obtained from the slate manufacturer.

Figures 10.1(a and b) give an indication of slating techniques. Figure 10.1(a) gives the typical construction for eaves and verge on a barge board. Note the special length slate at the eaves and the double batten. This second batten is needed to allow the eaves slate and the full first slate to be nailed independently – it is not possible to nail through the eaves slate from the slate above without causing damage. Each slate is supported on three battens, with the batten at the head of the slate far enough above the head to allow the next slate to be nailed directly into the batten. It is for this reason that very careful setting out of lap and gauge of the battens is vital with slate roofing.

Figure 10.1(b) shows one form of slate hip, and in addition to the notes above, the key points on the hip are the 'slate-and-a-half', which is shaped by the slate layer to the hip angle. The minimum width at the head of the slate should never be less than 50 mm, as this is its third support point on the batten. Before battens are fixed an additional 600 mm strip of underlay should be laid down the hip length with its centre on the line of the hip board from ridge to eaves. This gives additional water penetration protection, and because at the hip line slates can only be butted the cement mortar on which the hip tile is bedded is the main waterproofing for the roof at that point. The hip iron, which traditionally was a forged iron fitting (often with ornamental end) is screwed to the hip rafter with galvanised screws; its function is to hold the hip tiles in place not only while the mortar sets but later when mortar beds may fail – the iron stops the hip tiles from sliding off the roof. This hip iron is now usually made of stainless steel to prevent corrosion. The end of the hip tile is traditionally filled with pieces of slate bedded in the mortar.

For further information on slate roofing and construction, refer to the slate producer's technical literature (see the Bibliography for details).

CONCRETE INTERLOCKING TILES

Interlocking tiles in their most basic form were made of clay and used in Roman times. These consisted of a tile, semicircular in section, laid alternately like a channel and ridge; that is, the same tile is inverted to form the ridge, thus protecting the joint between the other two. Today, the more common interlocking tiles come in two main varieties: (i) the single interlocking tile where the interlocking shape is on the side of the tile; and (ii) the double interlocking tile which has an interlocking shape on both the side and head. Single interlocking tiles can be made of clay or concrete, but double interlocking tiles are only made of concrete. They are available in a wide range of colours and profiles, from traditional pantile shape through varying forms of architectural corrugation to imitation slate. Unlike slates and plain tiles, the interlocking tile does not require any overlap of the course below to protect the side butt joints from rain penetration. If you study the slate roof it can be seen that at no point on the roof is there a single layer of slates; there are at least two and at some points, three thicknesses to afford protection. With the interlocking tile, although half-bonding (i.e., the tiles laid with staggered joints) is traditional it is not essential for weather proofing. With no interlocking on the head of the tile, the single interlocking tile can have an adjustable head lap to cope with varying exposure conditions and is suitable over a wide range of pitch.

Figure 10.2(a and b) illustrate single lap tiles on a barge board verge, eaves and side abutment gutter detail construction. As there is little overlap of the tile, tile clips should be used at the batten as well as nails from tile to batten. It is also vital to use a tile clip at the verge to keep the tiles together and hold them to the roof via the batten. The cement bed at the verge should be seen as a weatherproof filler for the joint and not a means of 'sticking' the tiles to the roof.

Most tile manufacturers have a 'dry verge' product, and this consists of a matching colour plastic component which effectively clips the tiles to the verge. Details of such proprietary products can be found from the manufacturer's literature.

Double interlocking tiles have both side and head lap interlocking pattern and are consequently fixed in their course and lap dimension. They do, however, offer an extremely high resistance to wind and water penetration, and for that reason they can be used at much lower pitches than most other types of tile. As with the single lap tile, clipping is essential, particularly at the verges, as there is very little overlap and therefore very little weight from the tile above on the tile below, thus reducing its wind or indeed suction uplift resistance. Whilst the colour range is similar to that of the single interlocking tile, the interlocking head of the tile tends to limit the profiles to imitation slates and simple architectural corrugations. These tiles tend to be relatively large, and because there is minimal lap they can be laid very quickly; also, because the batten coursing tends to be relatively wide, that process is fast and economical in the use of batten.

Further information on tile manufacturers is provided in the Bibliography.

PLAIN AND PEG TILES
Clay Tiles

Other than thatch, these tiles must be the most traditional form of roofing used in the southern half of England. First used some seven centuries ago, the peg tile made of clay and then fired was a simple rectangle of about 150 mm wide × 250 mm long, pierced towards the top with two square holes through which were driven small wooden pegs. These pegs, often of oak, held the tiles in place over the batten, a function now provided by the clay 'nib'.

Clay plain tiles were developed over the years, these being slightly larger at 165 mm wide × 265 mm long, and included one or two nibs depending on the maker, and with two or three round holes for nails to securely fix the tiles to the batten. The peg tiles hitherto had simply been laid on the batten with only their weight to keep them in place, with the peg hooked over the batten. Being hand-made and with some distortion in firing, the tiles developed to have a slight curve towards the roof in both their width and length, thus ensuring that the meeting tiles discharged their water to the tile below. Being made of a natural clay, there was – and still is – a limited range of colours available, these being influenced by the type of clay in the locality of the tile maker. Colours traditionally range from 'brick' red to darker heather colours. Modern concrete versions of the plain tile are made to the same size and shape as stated above, but with a much wider range of colour choice.

Due to their small individual size, they are more 'flexible' to the roof shape. Indeed, it must be remembered that seven centuries ago roofs were constructed of selected branches rather than neatly sawn timbers, and consequently were not the flat, even slope of today's trussed rafter roofs. Traditional 'cottage' roofs also necessitated the roofing of small dormers with their associated ridges and valleys, and these small tiling units coped extremely well with both the unevenness of the roof and the inaccuracy and irregularity which resulted from the use of irregular timbers for the roof construction. The availability of special sized tiles – that is, half-width tiles, one-and-a-half-width tiles, special-shaped hip tiles and valley tiles – all helped to make this type of roof extremely popular. All of these 'special-shape' tiles are still available today.

Laying Plain Tiles

Pitch is critical, because with no side or head interlocking device (like slates) plain tiles rely on over-lapping each other both sideways (minimum one-third tile width lap) and head (minimum 65 mm) to provide weatherproofing. With their curved shapes and the irregularity of hand production, it is advisable to use a high-quality underlay and to ensure that it is well lapped and dressed at the eaves with good support. At hips, the use of the extra 600 mm width of underlay, as indicated in Figure 10.1(b), is essential, while good overlaps at valleys, as indicated in Figure 10.3(a), is strongly advisable. Permavent (a company mentioned above) has a system called 'Plain Easy', which allows some plain tiles to be laid down to a roof pitch of 17.5° (see the Bibliography for contact details).

Nailing to battens is even more essential with this relatively lightweight small tile to avoid wind displacement. Alloy nails (typically 38 mm long × 3.35 mm diameter) should be used, with two nails in each tile on the two eaves and ridge courses, and on each of the fifth course up the rafter. At verges and abutments every course should be nailed. As the specified battens for plain tiles are often only 25 mm thick (38 mm wide) for either a 400 mm or 600 mm spacing of the rafter, a high-quality timber is paramount because of the high number of nails required on certain battens between each rafter. For this reason it is strongly recommended that 38 mm-thick battens be used, especially as battens can often be delivered undersize. However, this extra thickness must be taken into account when setting out barge boards and fascias.

Check with the tile manufacturer for the exact fixing specification for their plain tile and recommended batten size, all to conform to BS5534:2014.

Mortar Bedding

Traditionally, lime mortar would be used for bedding clay products, and its inclusion in any mix should be used today. Unglazed clay is a breathable product, and can deteriorate relatively quickly if pure cement mortars are used. It is therefore strongly recommended that the mortar for bedding at any point on the roof should be made up of one part of lime to five parts of sharp sand with only one part of cement, which gives the mortar some added weather resistance where it is exposed. For concrete plain tiles, conventional mortar mixes of approximately one part cement to three parts of sand should be used.

The above applies to traditional building methods as is applicable to repair works, as well as new builds, to conform to BS 5534:2014. There is still the need to mechanically fix tiles, and specifications from the tile manufacturer should be obtained for the site location being worked on.

An exception to BS 5534 may be made if plain tiling is being carried out on a listed building, and roofing putties may not be allowed to be used, in favour of the lime mortar bedding.

Ventilation

Today's manufacturers of both clay and concrete plain tiles offer various forms of in-tile ventilators. Some, on a traditional clay tile, manufacture an in-roof ventilator comprised of three separate clay tiles, while others integrate plastic ventilation systems. Care must be taken in choosing such ventilators, which should of course be incorporated into any newly constructed roof whether on renovation or new build. The aesthetics of the ventilators may need to be discussed with your local Conservation Officer.

As always, consult the tile maker's literature, some details of which can be found in the Bibliography under 'Plain and Peg Tiles'.

Plain Tile Valley Detail

As shown on Figure 10.3(a), this is an open valley lined with lead, but in plain tiles there are alternative special valley tiles manufactured which allow the courses to run continuously around the valley. These valley tiles are simply laid in position on the battens and, being V-shaped, wedge themselves between the tiles of the two abutting roofs. There is also a 'laced' valley, which requires extremely precise setting-out of the tiling battens, allowing the tiles to overlap in a similar manner to the lacing of lines on shoes. For further information on these valleys, please refer to manufacturer's information.

Referring to Figure 10.3(a), the tiles at the valley are a cut-tile-and-a-half; this allows a good width of tile at the batten to allow two nails to fix the tile in position. Cutting a single tile could result in only one nail, which would not adequately secure the tile. The valley boards, usually of exterior grade plywood or treated timber, should be fitted between the rafters on a support batten and nailed to the side of the rafters. A tilt fillet cut specially to shape, supporting the felt to avoid ponding, carries the cut edge of the tile to maintain the correct pitch. It is usual to dress the lead in one continuous width from tile fillet down and into the valley gutter and up over the next fillet, securing with alloy nails. There should be at least 150 mm lap if the lead has to be joined in the length of the gutter.

A tile slip is then laid in line with the cut edge of the tiles and the tiles bedded on mortar to the mix described above.

This illustration also shows the difference between a peg tile with its wooden pegs in square holes, and the plain tile with its nibs and round holes for nails.

Plain Tile Sprocket Eaves Detail

A simple eaves detail with the fascia fixed directly to the foot of the rafter is illustrated in Figure 10.1(a). Figure 10.3(b) illustrates a sprocket eaves, an architectural feature often found on traditional cottages with a steep pitch, the sprocket helping to arrest the flow of water by lowering the pitch at the eaves. The difference in pitch can be seen at the angle A. In this particular illustration, exposed rafter feet are also illustrated, again a common feature with traditional cottages, which also leads to an exposed soffit that originally would have been left as the underside of the tile and batten. This soffit with its ventilator has a ventilation gap batten fitted on top of it to support the roofing underlay, and maintains the ventilation void. The eaves closure board secures the eaves against insect and bird ingress, and also controls the insulation.

The eaves tile, typically 190 mm long, should project well into the gutter, and careful setting out of the battens is necessary to give equal batten spacing either side of the pitch change, as indicated by the equal signs in the illustration. Care must be taken to ensure that the lower pitch of the sprocketed area is within the minimum recommended by the tile manufacturer.

ASPHALT SHINGLES

Bitumen felt roofing is usually associated with low cost non-dwelling buildings such as sheds, garages and workshops, and the resulting roof covering gives the appearance of a typically green (although other colours are available) flat sheet with little aesthetic appeal. Bitumen felt shingles, however, offer an economic, durable and aesthetic answer to all types of building, where a low weight roof covering is of advantage. These shingles are not the thin felt often associated with the rolls of felt available in builders' merchants, but are substantial, up to 3 mm thick, and comprised of a glass-fibre mat to give mechanical strength. They are pre-impregnated with bitumen to ensure a solid mat, and further bitumen is then added to ensure stability and resistance to temperature fluctuations. Coloured ceramic granules are then bedded in the surface, the range of colours matching closely those of traditional clay or concrete roofing. The underside is coated with silicone sand and a special thermosetting adhesive is then applied to the top surface, which bonds the tiles together when laid. The overall effect is that of a tiled roof, with each shingle being divided into three, four, or five tiles, the smaller number mimicking a slate roof, and the larger number representing a plain tiled roof.

Being bonded to the deck of the roof, the shingles can be laid to a very low pitch, typically down to 14°, but even pitches lower than that are possible with special torch-on-underlays applied to the deck before the shingles are laid. It can be seen, therefore, that this is a very versatile product, coping with valleys, hips, ridges and abutments with ease, and can easily be undertaken by a competent DIY enthusiast.

Structure

The shingles require a deck of usually exterior plywood to be laid over the rafters, typically a minimum of 12 mm thick. The verges and eaves should have a metal trim fitted (available from the manufacturers), with each length of trim (typically 3 m long) overlapping by 75 mm and well nailed to the deck. It is advisable to apply the special bitumen mastic adhesive over the nail heads for additional protection and where the sections lap. The eaves course of tiles is cut from a standard shingle and is laid on bitumen mastic applied on the metal eaves trim with the pre-applied shingle mastic pointing down to the eaves. Nails should not be fixed through the pre-applied adhesive, but five nails per-shingle should be fitted above the mastic line through the eaves trim and into the roof deck. The nails to be used are galvanised clout nails 8 mm longer than the thickness of the deck. This ensures that the nail fully pierces the deck; the fibres grip the nail and resist it springing out. This fact must be considered carefully if it is intended to leave the underside of the deck open to view inside the building, perhaps a garage or workshop. The next course of tiles is then laid conventionally with the decorative tile surface down to the eaves and again nailed through the first eaves course and above the adhesive line. Subsequent courses are laid to half-lap, thus imitating conventional tiling.

In cold weather it may be necessary to gently warm the shingles before applying them to the roof, and furthermore the pre-applied adhesive may need warming to perform a satisfactory bond. Failing that, additional mastic gun applied bitumen adhesive (supplied by the manufacturer) can be added to ensure that the tiles are satisfactorily bonded to the course below.

The ridge is formed by cutting the tiles from their shingle, shaping them as indicated in Figure 10.5, bonding and nailing them over the ridge. If the building is located in a particularly windy area, then the lap of the ridge should be downwind to avoid any possibility of the ridge tiles lifting.

Ventilation
By applying the felt shingles to the roof, with or without an underlay, a vapour-impermeable layer has been created, and ventilation of the roof space is essential. Manufacturers provide in-roof ventilators, and ventilation should be provided at eaves and ridge or if the building has no ceiling; then, rather than penetrating the ridge, ventilation can be provided high in the gable ends.

For further information refer to the Bibliography under the heading 'Asphalt Shingles'.

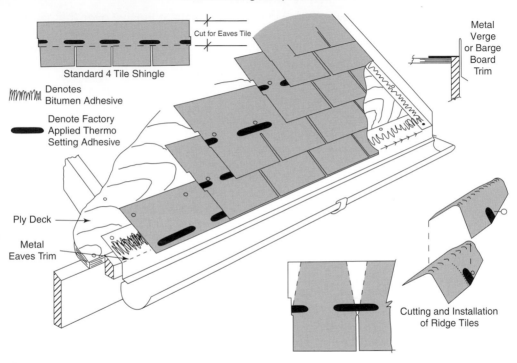

Goss's roofing ready reckoner

Cut for Eaves Tile

Standard 4 Tile Shingle

Denotes Bitumen Adhesive

Denote Factory Applied Thermo Setting Adhesive

Metal Verge or Barge Board Trim

Ply Deck

Metal Eaves Trim

Cutting and Installation of Ridge Tiles

Figure 10.5 Asphalt shingle roofing details.

METAL TILES

Like the asphalt tiles in the previous section, metal tiles are very common in Europe, but gain limited favour in the UK. In a similar way to the asphalt shingle, the metal tiles are available in strips comprising a number of tiles of various corrugated shapes pressed into the metal. The metal is then coated with rust-resisting compounds, and the top surface coloured and textured with further applications to give the appearance of traditional tiling. Again, the product offers a very lightweight alternative to traditional tiling, can be laid to a minimum pitch of 10°, but unlike the asphalt shingle, does not require a solid deck on which to be laid. Instead, they require a conventional rafter roof (or indeed a sarking finished roof) and battens, or in the case of the sarking roof, battens and counter battens. The battens to which the tiles are fixed are generally 50 × 50 mm treated softwood battens, the tiles being fixed with proprietary 50 mm-long nails. As some of the nails are left exposed, these need to be coated with the special touch-up paint available from the tile manufacturers.

Unlike other roofing, metal tiles are laid with the first course at the top of the roof, with the head being nailed to the batten. The next course down is then fitted by lifting the upper course, sliding the tile under, and nailing through the lower edge of the upper tile at one of the high points on the corrugation to avoid water ingress. Verges or bargeboard cover profiles, abutment weathering and flashings, and ridges are all provided by the manufacturer as matching pressings. It should be noted, however, that some special equipment is required to both cut the tiles and bend the tiles at hip junctions and abutments. Cutting can be done using tin snips (but this leaves a messy and not very safe edge) or with sheet metal cutters. All cut edges should be treated with the maker's special touch-up products.

Ventilation

Again, being a relatively thin metal roof, and although some air passage is possible though the tile lap, it can only be considered minimal and therefore the roof is 'impermeable'. Adequate ventilation must be provided and special matching fittings are available from the manufacturers. These include vents for installation at the fascia, within the roof rafter length, and at both the ridge and the hip situation.

Figure 10.6 illustrates typical steel tile roof covering details (for more details, see the Bibliography).

LEAD SUBSTITUTES

Metal theft from roofs during recent years has caused much damage by allowing water ingress into the roof and, in some cases, the buildings below. Lead is a durable soft, malleable metal, which makes it relatively easy to dress to shape to form flashings, soakers, valleys and damp-proof courses, apart from being a very good roofing material for very low-pitch or even flat roofs.

To avoid the problem of lead theft, not only from high-profile buildings such as churches but also from housing, there has been a wide range of substitute materials developed for locations such as those shown in Figures 10.2 and 10.3. One of the great advantages of lead has been its long durability of hundreds of years, provided that it has been fixed correctly. The negatives in fixing lead is the health risk in handling the material, and the careful avoidance of fixing with metal fixings, which can cause bi-metallic corrosion of fixing on lead (copper and lead were used in older acid batteries). The high coefficient of expansion of lead limits the length of a flashing or valley without making a joint to provide for this expansion and contraction.

The advantages of these lead substitutes claimed by many of the manufacturers are as follows:

- No scrap value means no risk of theft.
- Cheaper than lead.
- Simpler to install.
- No health and safety issues; however, as heating can be necessary on installation in cold weather and knives are involved for cutting, their use is not 'risk free'. Rather, the fact, the claim is based on the non-toxic nature of the product, whereas lead is poisonous.

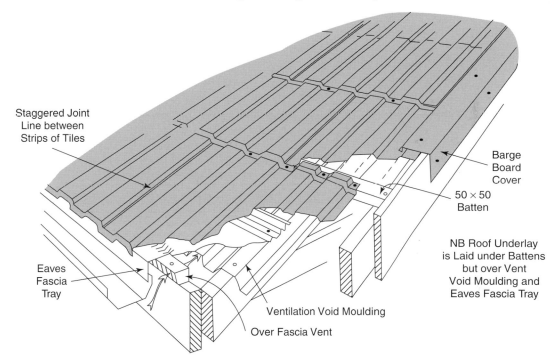

Staggered Joint
Line between
Strips of Tiles

Barge
Board
Cover

50×50
Batten

NB Roof Underlay
is Laid under Battens
but over Vent
Void Moulding and
Eaves Fascia Tray

Eaves
Fascia
Tray

Ventilation Void Moulding

Over Fascia Vent

Figure 10.6   Metal tiles – typical details.

- The products are recyclable.
- They are lighter in weight.
- Most are BBA certified and some issue a guarantee for 25 years.
- Available in range of tiles matching colours, Terracotta, grey (to look like lead) and black.

The products range from simple bituminous felt (relatively short life and difficult to dress to shape without cracking) to sophisticated multilayer products which include metal mesh inner layers, and gradual finishes. One such product is made by Ubbink (UK) Ltd and is sold under the tradename of 'Ubiflex'. Whilst these products can be dressed to a degree, they do rely on mastics to seal joints, except in the torch-on (i.e., heat-applied) types. One product, 'Wakaflex' (produced by Haus Profi)' claims to be able to be dressed to shape and has a peel-off backing that allows it to adhere to the building material to which it is being fixed. This is not a bitumen-based product but rather is a polymer with an aluminium mesh layer with butyl adhesive, but only a 10-year guarantee is offered. It is available in four colours similar to that above (for further contact details, see the Bibliography).

11   SOLAR PANELS

Solar panels for both electricity production (photovoltaic; PV) and thermal (hot water) have increased in popularity over the past few years, encouraged by the various government schemes for so-called microgeneration, which pay the owner of the system for electricity fed into the National Grid (PV systems) or for the equivalent power produced to heat the water.

It is not the object of this book to explore the economics of solar-generated power, that will have many variables and be specific to each individual installation. For those who wish to explore this further, impartial information can be found on The Energy Saving Trust website: www.energy savingtrust.org.uk/domestic/content/solar-water-heating and www.energysavingtrust.org.uk/dom estic/content/solar-photovoltaicpanels

Suffice to say, savings on energy bills can be significant, and in present times of low interest rates, solar energy can provide a high rate of return on investment. With the increasing efficiency year on year of the systems (especially PV) and the reducing cost of panels, converting the sun's energy to power can only become more financially attractive.

PLANNING AND ROOF SUITABILITY
Both types of panel will work in bright daylight, though PV and thermal panels are both at their most efficient in direct sunlight. For this reason, a south-facing roof will produce the highest output, although south-east and south-west facing systems can be viably productive. The pitch of

Goss's Roofing Ready Reckoner: From Timberwork to Tiles, Fifth Edition. C. N. Mindham.
© 2016 John Wiley & Sons, Ltd. Published 2016 by John Wiley & Sons, Ltd.

the roof also has an effect on efficiency, with about 30° being the optimum. Possible shading by other buildings or trees will reduce efficiency.

Each PV panel is approx 1.0 m wide × 2.0 m high (this varies slightly from different manufacturers); smaller panels of high output are available but at increased costs. The Government limits the microgeneration on domestic homes to an output of 4 kW, which usually means an array of 16 panels for PV, and two or three panels for thermal hot water. The PV system can have a significant visual impact on a dwelling, and therefore planning permission may be an issue. Both, solar PV and Thermal are seen as 'permitted development' under current planning laws, except on listed buildings, buildings in conservation areas and other geographical locations (see more on this in Chapter 12). Check with the local planning authority before proceeding and, if permitted, obtain a certificate from them known as the 'Certificate of Lawfulness'. This could be useful in providing proof of legality of the installation if and when the property is sold. There is a small cost for the certificate.

Building regulations also apply, but an accredited installer will deal with this aspect unless structural work is found to be necessary during the viability survey. The survey for suitability and the installation, including all plumbing and/or electrical work, should be entrusted to a MCS (Microgeneration Certificate Scheme) Contractor. This will ensure that all requirements of the Government-approved payment system, known as the 'feed-in tariff', are met.

ROOF CONSTRUCTION

Most trussed rafter-constructed roofs will prove adequate, but a trained eye – probably that of the viability surveyor – may suggest a more detailed structural assessment for older roof constructions such as traditional cut roofs and truss and purlin roofs. Modern cut roofs will have been subject to building regulation approval and are likely to be more than adequate to carry the load of the panels. The weight of the panels is not significant when compared to the weight of the roof itself and the tile or slate covering. Snow and wind loads will also not be substantially changed by panel application. Solar thermal panels are the heavier of the two types, at approx 19 kg/m^2 compared to 12 kg/m^2 for PV panels.

Some issues have been reported concerning the adequacy of the fixing of the panels to the roof due to wind loads imposed on them, particularly in areas of high wind exposure. The Building Research Establishment (BRE) National Solar Centre, has published two relevant digests as follows:

DG489 (rev. 2014): 'Wind loads on roof-mounted photovoltaic and solar thermal systems'.

DG495: 'Mechanical installation of roof-mounted photovoltaic systems' (see the Bibliography for contact details for both digests).

There is also an interesting and informative report prepared by The Scottish Government, Directorate for the Built Environment Building Standards Division (March 2010) entitled 'Risk Assessment of structural impact on buildings of solar hot water collectors and PV tiles and panels – final report'.

FIXING THE PANELS
Installation commences with setting-out where the array is to be located and the removal of tiles where fixing brackets are to be fixed; these brackets are fixed to the rafter through the tiling underlay with screws. To these brackets, after tiles have been replaced, are fitted the longitudinal rails to which the panels are bolted. The brackets are simply fitted with screws to the rafter. The screws should be of stainless steel or, as a minimum, galvanised steel. The maximum size of the screws to conform to design standards for timber construction for a 35 mm-wide trussed rafter is only 3 mm diameter. Many solar systems are fitted with up to 6 mm- and even 8 mm-diameter screws which require a thicker timber than the standard trussed rafter. However, this may not be adequate to resist wind uplift on the panels. A thicker screw could split the rafter and therefore reduce its ability to generate the full strength of fixings. The screws should not be fixed in the area of the truss connector plate. The workman fixing the bracket is unable to see the exact location of the rafter

or the truss connector plates through the underlay, and there is a risk of catching the edge of the rafter, making the screw fixing inefficient or disturbing the trussed rafter structural connector plate.

For the above reasons the trussed rafter industry is reviewing the recommendations for fixing solar panels to their product. The BRE Digests, as mentioned above, offer a method of calculating wind uplift on panels throughout the country, taking into account the exposure (i.e., height above sea level, topography and location). From these loads a structural engineer could calculate the precise number and type of fixings required to resist the load. For a trussed rafter roof, the engineers at the manufacturer (if identifiable; see Fig. 8.1) should be consulted.

Most PV and thermal panel system suppliers have their own fixing systems, but most pay little attention to the roof structure itself. In researching for this chapter, one company – Klober Ltd – does produce a whole range of products for the better installation of solar panels and continued waterproofing of the roof. The range extends from a universal panel mounting system with its own weatherproof flashing unit to special flashings allowing cables (PV) and pipes (solar thermal) to pass through the roof (for details, see the Bibliography).

POSSIBLE IMPROVED FIXING
To accommodate the 6 mm- or 8 mm-diameter fixing screw a thicker piece of timber could be fixed to the side of the trussed rafter or common rafter on a traditional roof. This need only be fixed locally in the location of the fixing bracket, and could be screwed or bolted (not nailed) to the rafter. A qualified engineer or the trussed rafter manufacturer could advise on the number and size of the screw or bolt to be used. An alternative – and probably easiest to install – is a light metal (galvanised steel or stainless steel) strap that would pass through the tile underlay and again be fixed to the side of the trussed rafter or rafter. This would probably be nailed satisfactorily with special nails supplied by the manufacturer, as many other special metal fixings are fixed to timber structures, for example, truss clips to wall plates and joint hangers to trimmer joists. These suggestions are illustrated in Figures 11.1 and 11.2.

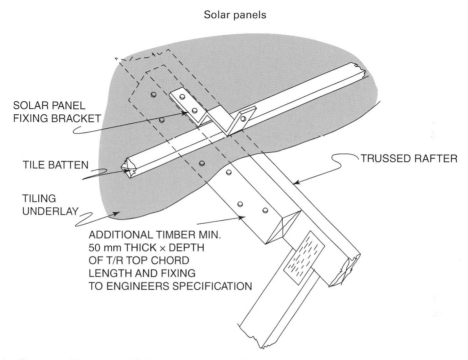

Solar panels

SOLAR PANEL
FIXING BRACKET

TILE BATTEN

TILING
UNDERLAY

TRUSSED RAFTER

ADDITIONAL TIMBER MIN.
50 mm THICK × DEPTH
OF T/R TOP CHORD
LENGTH AND FIXING
TO ENGINEERS SPECIFICATION

Figure 11.1 Proposed improved fixing to trussed rafters.

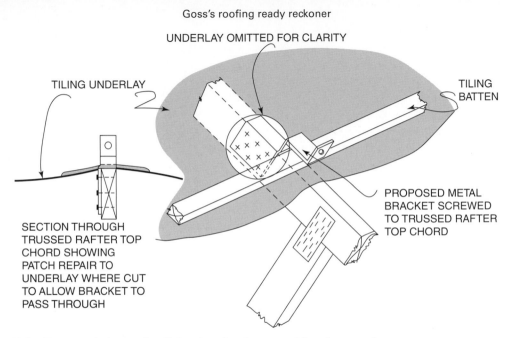

Goss's roofing ready reckoner

UNDERLAY OMITTED FOR CLARITY

TILING UNDERLAY

TILING
BATTEN

PROPOSED METAL
BRACKET SCREWED
TO TRUSSED RAFTER
TOP CHORD

SECTION THROUGH
TRUSSED RAFTER TOP
CHORD SHOWING
PATCH REPAIR TO
UNDERLAY WHERE CUT
TO ALLOW BRACKET TO
PASS THROUGH

Figure 11.2 Proposed alternative fixing bracket for use with solar panels.

VARIATIONS ON PANEL DESIGN

All of the above has assumed the panels are installed above the roof covering of tiles or slate, leaving an air gap of 25–50 mm for air flow. 'In-roof' panel systems exist which take the place of the tile or slates and have a waterproof tray fitted to the roof as well as a system of top, side and bottom flashings to maintain water and wind proofing. The flashing systems are similar to those used for roof windows.

Another variation is strips of solar tiles manufactured to match conventional tiles, thus blending into the roof's overall appearance. These strips are fixed using the same method as is used for the tiles into which they are inserted, that is, direct to the tiling battens with either nails or screws (see the Bibliography for further information under metal roof tiles).

12 SHEDS AND OUTBUILDINGS

PERMITTED DEVELOPMENT
Although the planning rules have been relaxed over the last few years to allow more varied types of outbuildings to be constructed without planning approval, it is wise to run your ideas and proposals for your outbuilding past your local planning officer. It would be wise to have prepared a drawing showing its overall size (there are limits within permitted development) and show the proposed building location both to your house and the boundary of your property. The planning officer will also be able to advise you if you should consult the building control department of your local council.

TYPES OF PERMITTED BUILDING
Apart from size limitation, the type and use of the building in broad terms, is a building which you do not sleep in. This therefore allows:

- Sheds for general storage.
- Greenhouses.
- Kennels (unless intended for use as a commercial enterprise).
- Home offices.
- Workshops.

Goss's Roofing Ready Reckoner: From Timberwork to Tiles, Fifth Edition. C. N. Mindham.
© 2016 John Wiley & Sons, Ltd. Published 2016 by John Wiley & Sons, Ltd.

- Garages.
- Games rooms.
- Summer houses.

A very useful website on this matter is to be found at 'Planning Control – Planning Portal – Outbuildings':

- No outbuilding must be forward of the main dwelling, that is, in front of the building line.
- Must be SINGLE STOREY, maximum eaves height 2.5 m and overall height not exceeding 4 m for a dual-pitch roof, or not exceeding 3 m for any other roof (see Fig. 12.1).
- The maximum height must not exceed 2.5 m if within 2 m of the boundary of the curtilage of the dwelling house.
- No verandas, balconies or raised platforms are allowed.
- Not more than half the area of the land around the original house. This means as the house was originally built, or as it stood before 1948. This could be a tricky rule to interpret, especially if extensions have been made to the main dwelling. Consult your planning office.

Separate rules apply to dwellings in:

- National Parks.
- The Broads.
- Areas designated as being of outstanding national beauty.
- World heritage sites.
- Listed buildings.

If in doubt check with your local planning office.

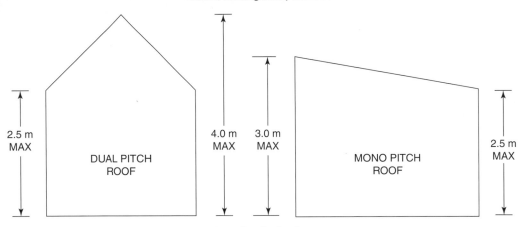

Figure 12.1 Permitted building cross-section size limitation.

There is also some technical guidance on the above website under 'Technical Guidance' for permitted developments. This is good for looking at the rules for extensions to dwelling which have not been mentioned above and although may be 'permitted' under planning, will definitely require building regulation approval.

Building regulations may be required for outbuildings, for such matters as electrical work (it is advisable to employ an approved electrician anyway, in order that they will supply a certificate for the work, which could be very valuable if the property was sold). Also, fire resisting regulations apply to the outbuildings construction if close to a boundary, particularly a boundary to neighbouring property. Check with the building control office.

DESIGN

The plan shape, overall size and the height of the building will have to be within the above planning constraints. So too must its location within the boundary. The materials used for the construction of the walls will depend to an extent on how important it is to harmonize with existing building and, to a degree, its intended life span. It may only be intended as a short-term structure. Depending on its fire-resisting requirements (as noted above) it may be built of brick to match the existing dwelling, blockwork, timber frame with timber or fire-resisting cladding, plaster rendering on block or a timber frame and lath structure.

This book, however, is concerned with the roof structure and roof covering.

THE STRUCTURE

Because of the constraints on size imposed by planning regulations, the roof structure is not going to be too complex in terms of designing for adequate strength.

Having decided on the roof shape – that is, mono-pitch, dual-pitch, and so on – consideration should be given to the following points:

- The intended lifespan of the building; that is, the durability of the roof covering.
- The fire resistance of the roof covering, if any is required by building control.
- The weight of the proposed roof covering; this will affect the design of the structure.
- Is the roof void to be used as storage? This also will affect the design of the structure.
- Rainwater drainage and disposal. Too many sheds have their life dramatically reduced because rain water run-off is allowed to drain down the walls. Gutters are very important but they must be discharged to a good-sized water butt, making use of the water, or to a suitable soak-away!

THE MONO-PITCH ROOF

This is the simplest form of roof and needs little design of the structure. Assuming timber rafters will be placed from ridge down to eaves, the timber size can be calculated from the tables in Chapter 3. However, if a lightweight roof (rather than conventional clay or concrete tiles) is to be used, it would be safe to use the design spacing from the tables for a 400 mm rafter spacing and extend this to 600 mm for the building. This would cope comfortably for metal tile roofing or metal sheet roofing on battens over the rafters, or felt sheet or felt tile strips on a plywood or OSB (Oriented Strand Board) fixed to the rafters. If conventional tiles or slates are to be used, size the timber as stated in the tables.

The finishing at ridge, eaves and verge will be, to an extent, in line with some of the details illustrated in earlier chapters or to the roofing material producers specific instructions. Ensure that the correct fixings are used as specified by the roofing material manufacturer.

THE DUAL-PITCH ROOF

In its most simple form, the dual-pitch roof delivers the load from the tiles down the rafters on to the eaves wall plate of the building. This load is more complex and results in some sideways push outwards on the wall plate and the wall of the building below. Consequently, the wall must be stiff enough in its length to resist that load without bending. If it does bend outwards this will result in either the collapse of the wall or the typically seen bending of a timber-framed wall, accompanied by a sagging roof!

To overcome this situation there are two solutions: (i) fix the rafters to a ridge and mid-span of rafter purlin (see purlin in Figs 2.1 and 3.1) or (ii) fit a collar across the rafters (see Fig. 12.2). The lower down the rafter the collar is fitted, the more stable will be the roof structure. The temptation with the collar solution and where the roof void above is left open, is to use the collars as a storage rack! If so, ensure they are adequate to take the load. Design for the collar size can be assessed from the design table for ceiling joists and the heavier loading (i.e., 'more than 0.25 kNn/m^2 but not

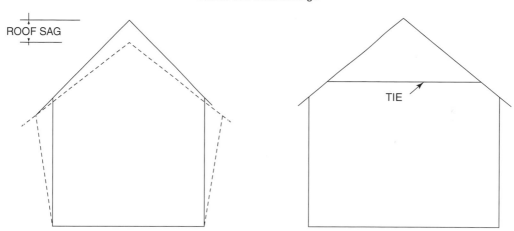

Figure 12.2 Stability effect of rafter tie.

more than 0.5 kN/m^2). It is recommended that the fixing of collar to rafter is by a 10 mm-diameter bolt, nut and 50 mm square washers (see Fig. 12.3).

Timber member design for either of these options can be taken from the tables in Chapter 3, again taking the roof finish loading into account. For a concrete or clay tiled roof use the table sizes as published, but for a lightweight roof (as listed above) stretch the size for 400 mm spacing to 600 mm, as with the mono roof.

A note on the purlin roof option, as this puts the majority of the roof load on the gable ends, and therefore they must be strong enough to take the load. Fitting rafters to a purlin or wall plate should have a birdsmouth cut to ensure the rafter does not 'slide' down (see the illustration of a birdsmouth in Fig. 2.3). The strongest timber-to-timber connection is by galvanised connector plates; these are available at all builders merchants and are fixed with small special nails available with the plates. The use of large nails can easily split the timber and does not always result in a strong joint. Typical metal connections are illustrated in Figure 7.1. Figure 12.3 shows a mixture of construction and roof covering alternatives, as discussed above and below.

ROOF COVERINGS
The decision on the roof covering has been briefly considered earlier in this chapter. Cost also must be considered, and this will be influenced by the intended use of the building and its life expectancy.

Options:

- Conventional concrete or clay tiles on underlay and battens on rafters (see Figs 10.1, 10.2 and 10.3).
- Lightweight metal tile strips on battens underlay and rafters (see Fig. 10.6).
- Heavy-duty bitumen felt tiles fixed on ply or OSB boarding fixed to rafters (see Fig. 10.5).
- Corrugated metal sheet fixed to battens on underlay on rafters or on purlins (see Fig. 12.3).
- Corrugated bitumen sheets fixed to battens on rafters or purlins.

NB: On either of these sheet covering options, it may pay to use a ply or OSB over the rafters if it is intended to have a 'ceiling' in the building at that position.

For all but the traditional tiled roof the manufacturer's fixing instructions must be obtained and studied BEFORE starting the operation. Also, with all of these non-traditional roof finishes there are special ridge, eaves, verge and fixings available. If used as specified, these will provide a sound roof against both rain and, very importantly for lightweight structures, against wind damage.

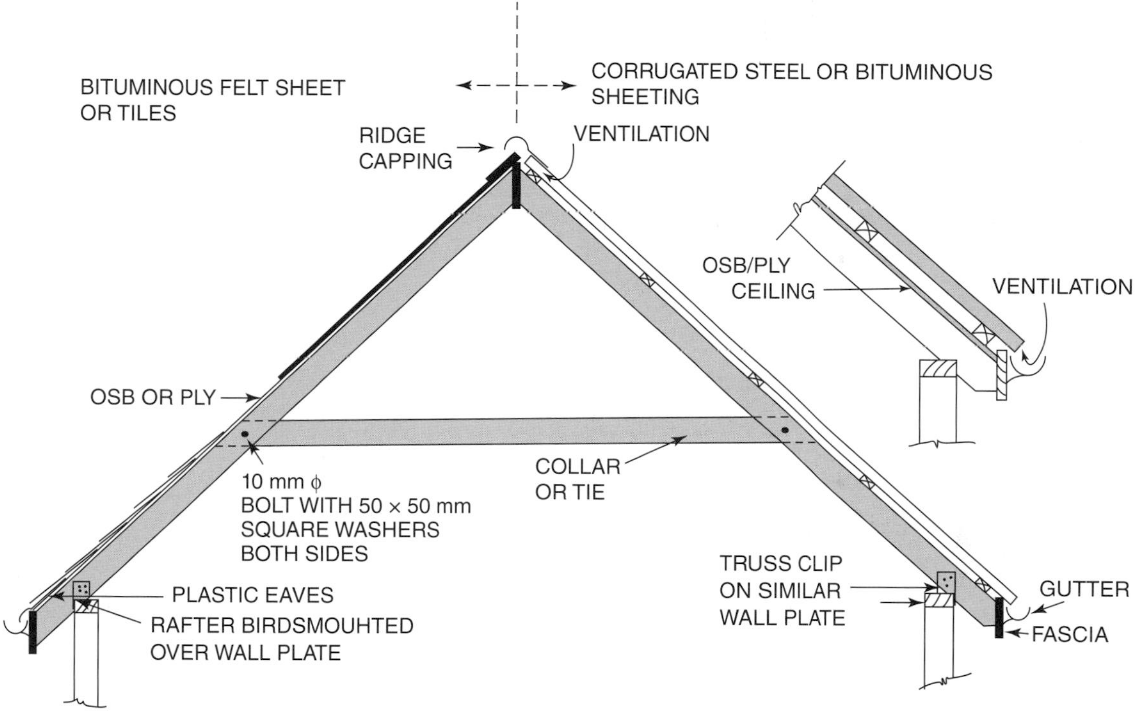

Figure 12.3 Typical outbuilding roof construction.

MAKE YOUR NEW BUILDING SAVE MONEY!

We have looked at many types of roof covering, but there is one which could save money in the long term. Solar panels are now available at affordable prices and could be fitted to a simple waterproof underlayer, thus saving on tiles or other coverings. There are even PV panel systems which can be coupled together with weatherproof flashings which could be used without any other roof covering – that is, they could be your tiles! More information on PV panels is provided in Chapter 11, but if you do wish to explore this option the roof slope size would have to be designed around the panel size available.

13   TOOLS AND EQUIPMENT

The roofing carpenter will need a number of tools and pieces of equipment to satisfactorily obtain information from drawings, mark the timber, set out the roof on the wall plate, cut the timber, and check the completed roof for line, level and plumb. A conventional pencil or pen may be needed for paper calculations, but the true carpenter's pencil should be used for marking timber.

OBTAINING INFORMATION FROM THE DRAWING

(a) Scale rule: only to be used if dimensions are not clearly shown on the drawing.

(b) Protractor: to measure the angle of the roof, but again only if the angle is not written on the drawing.

(c) To mark out the length and angles to be cut on the timbers: steel measuring tape of minimum 5 m in length now available with a digital display to remove the possible error of misreading.

(d) A bevel: a simple carpenter's adjustable bevel is adequate, this being set to the protractor to obtain cutting angles.

(e) Alternatively, a combination square with centre head and built-in protractor is more versatile.

(f) Alternatively, a digital bevel may be used for instant visual display of the setting angles required.

(g) A traditional roofing square can be used if the carpenter is trained in the use of this particular tool.

Goss's Roofing Ready Reckoner: From Timberwork to Tiles, Fifth Edition. C. N. Mindham.
© 2016 John Wiley & Sons, Ltd. Published 2016 by John Wiley & Sons, Ltd.

TO CUT THE ROOF

(h) Hand saw.

(i) Mains or 110 V electric hand saw. Cordless powered hand saws are available, but a mains power source is still required to recharge batteries.

(j) A compound angle mitre saw. This is an electrically powered saw designed especially to cut angles on timbers – check that the saw is large enough to cope with the length of the cut required; a 300 mm-diameter saw should be adequate. This type of saw can tilt in both planes and therefore be set to cut compound angles – that is, the ridge and edge bevels on hip jack rafters – in one operation. This type of saw is invariably fixed to a bench or stand, and therefore support will be needed for long timbers to be cut. This support should be fitted with a sliding stop system to allow quick repeat lengths to be cut without remeasuring. **NB**: For safety, all power tools on site should be 110 V – for home use a 240 V saw may be used only in conjunction with a power protection plug adapter which will protect against accidentally cut cables and faulty wiring, possibly causing electric shock to the operator.

SETTING UP THE ROOF STRUCTURE

(k) A steel tape at least as long as the roof wall plate itself.

(l) A good level; by this we mean a good quality level at least 900 mm long with two spirit levels, one for horizontal use and one for vertical use for plumbing timbers.

(m) Alternatively, a good-quality level as above but with digital angle readout would be useful to check work on the roof as it proceeds.

(n) To check and set level and verticals over longer distances (i.e., beyond the 900 mm of the levels itself), a laser level could be used. A wide range of laser beam projecting levels are now available giving an accurate beam projection of up to 50 m. The sophisticated rotary laser levels will give both horizontal and vertical beam projection, but it must be remembered that these are accurate only if they are mounted on a stable structure.

ROOF COVERINGS

Whilst many of the pieces of equipment and tools mentioned above will be required in the setting-out, the fitting of the underlay, and the cutting, fitting and nailing of the roof tiling battens, the number of individual and specialist tools required for the various types of roof covering are almost too numerous to mention. Certainly, some form of tile cutting device will be required and this may well be of a power saw type, and there will clearly be mortar to mix, with the requirement for some accurate method of measuring the various proportions of sand, lime and cement, and so on.

Cutting holes in slates will need a special slate holing tool, and whilst the cutting of asphalt shingles requires no more than tin snips or indeed a good sharp knife, metal tiles should be cut with a powered disc cutter, and will require specialist bending equipment available from the manufacturers.

If lead is to be used on the roof, then a range of lead dressing tools will be required, traditionally in the form of wooden mallets and beaters which 'dress' the lead to shape.

Never try to save money by not buying the correct tool for the job. The correct tool, often developed over many centuries by skilled craftsmen, will make the job both easier, faster, and give a more professional result.

14 HEALTH & SAFETY CONSIDERATIONS

ACCESS TO THE ROOF

For anything other than minor maintenance to the roof, a full scaffold or properly designed system of scaffold towers should be constructed in the area on which work is to be carried out. If this is the entire building, then scaffolding will be needed around the entire building. For professional builders, this is likely to be undertaken by a professional scaffolding company, who will visit the site if necessary and advise on the scaffolding required, deliver and erect and adjust them from time to time, as may be necessary for safe access to the work. Whilst the professional construction industry is covered by the many rules and regulations including the 'Working at Height' regulations which came into force on the 6th April 2005, the DIY builder and the self-builder who is essentially carrying out the work himself and not employing others, is not covered by the regulations. It is still wise to have qualified professionals build the scaffolding and access platforms that may be required, if only for insurance reasons should an accident occur and a claim need to be made on personal insurance.

Goss's Roofing Ready Reckoner: From Timberwork to Tiles, Fifth Edition. C. N. Mindham.
© 2016 John Wiley & Sons, Ltd. Published 2016 by John Wiley & Sons, Ltd.

BASIC PRINCIPLES

- Assess the risks involved in each operation.
- Plan a safe method of working for each operation.
- Decide what personal protection equipment will be required for each operation.

RESTORATION AND RENOVATION OF EXISTING ROOF STRUCTURES

If there is any doubt about the structural soundness of the roof structure itself, this should be carefully examined by a competent person to ensure that it will be safe to work on, and below.

Decide on the access required, taking into account any additional loads on the scaffolding which may occur from the temporary storage of carefully removed roof slates or tiles. If the roof covering material is to be disposed of, ensure that there is a rubbish chute discharging directly into a skip; do not simply throw the tiles to the ground.

Roof structures generally present a problem of access from inside the building, assuming that a full scaffold is to be erected on the outside of the building including the gable ends. It is quite likely, on a restoration project, that the ceilings will be removed even if the ceiling joists are satisfactory and being left in place. Whether the ceiling is lath and plaster or indeed a sheet material, it is best removed from above rather than below because of the fragments, falling elements of the ceiling, and the usual copious amounts of dust. If working from below, then full personal protection equipment should be used, including a safety helmet, goggles rather than glasses, the appropriate respiratory mask, stout gloves and steel toe and soled shoes or boots. It may also be worthwhile using one of the disposable all-in-one overalls.

When removing old timber, the rusty nails with which they are connected are a major hazard and should be treated with great respect. Good stout gloves are a necessity for such work, and all nails should be removed where possible or at least bent down and hammered well into the timber for safety. It is likely that replacement timbers will be preservative-treated, the preservative itself giving rise to a further Health & Safety hazard which must be addressed. All treated timbers should be delivered dry, but recutting and retreating with preservative on site necessitates handling with liquid-proof gloves, wearing safety glasses to prevent splashing in the eyes, and a waterproof apron and boots. From an environmental safety viewpoint, the timber treated offcuts should be properly disposed of at a local authority waste disposal centre. They should not be dumped with general rubbish, nor should they be burnt on site.

NEWLY CONSTRUCTED ROOFS

Again, proper scaffolding should be provided all around the perimeter of the roof. If working on a single-storey building, or on a building where the floor immediately below the ceiling level which is being worked on has been decked out, then 'soft landing bags' should be used to mitigate the consequences of a fall. Erecting a new roof is a notoriously hazardous part of any building operation, and this applies to traditional roof construction where the support structure is progressively put in place – that is, ceiling binders, purlins, ridge and so on – but particularly to trussed rafter construction where both the hoisting of the prefabricated component, its safe handling whilst being manoeuvred into position, and its temporary fixing in position all pose their own problems. On large-scale house building sites it is now becoming more common for the roof structure to be built on the ground in its entirety, including in some instances felt and battens, and then hoisted into position by crane. This clearly eliminates a number of the hazards in erecting the roof structure.

THE ROOF COVERING

The application of the roof underlay and battens clearly has risks of falling through the roof structure itself, and this can be greatly reduced by laying scaffold boards safely over the ceiling joists of the roof being worked on. When working on an underlay and battened roof, great care must be taken to walk up and down the roof only on the line of the rafter.

All roof coverings are heavy, some considerably more than others. Whilst steel tiles are relatively light for the area they cover, they are likely to be loaded onto the roof in packs of several kilos each. With plain tiles, on the other hand, whilst individually not heavy the sheer quantity required for the roof covering will add almost 1 ton to an average sized house or bungalow. Consideration should be given to exactly how these are going to be lifted to the roof, and if they are going to be temporarily stacked on the scaffolding, which will have to be designed to carry their weight. Great care must be taken to ensure that the roof is loaded equally on both sides of the ridge when loading-out the roof with its tiles or slates. Care must also be taken not to place large stacks of tiles on the roof, but to spread the stacks evenly immediately above a rafter and equally up the rafter length, thus mimicking the load which the rafter will eventually take. Temporary overloading of the rafter can cause buckling if it is not adequately restrained by diagonal bracing, and because the buckling can be passed from one rafter to another via the tile battens, distortion at the gable end can occur.

CONCLUSION

The above is by no means intended as an exhaustive list for the safe construction of roofs and their coverings. The number of operations involved from the bedding of the wall plate to the placing of the last tile or slate on the roof involves:

• Sawing of timber by hand saw or power saw.
• Use of preservative treatments.
• Application of nails by hammering.
• Drilling with hand or power drills for bolts.
• Lifting of sometimes heavy steel or timber components.
• Power cutting of clay, concrete and metal tiles.
• Use of sharp knives and tin snips for asphalt shingles.

- Constant handling risks whilst laying grit covered concrete tiles.
- Sharp edges of freshly cut slates and tiles.
- Use of cements and mixing of mortar for bedding the roof covering.
- Finally, the possible use of plasticisers and accelerators in the mortar.

Carefully consider all aspects of Health & Safety during construction of the project; a serious accident could mean that you will not complete it.

Safe building!

BIBLIOGRAPHY

Listed below are details of the publications and products mentioned in this book. For alternative products it is advisable to use the internet, or visit local builders and timber merchants.

ATTIC ROOF TRUSSES AND LOFT CONVERSION COMPONENTS

For list of trussed rafter manufacturers:
Trussed Rafter Association
The Building Centre, 26 Store Street, London, WC1E 7BT
Tel.: 020 320 0032
www.tra.org.uk
Email: info@tra.org.uk

For details of "JJI-Loft":
James Jones and Sons Ltd,
Timber Systems Division, Greshop Industrial Estate, Forres, Scotland, IV36 2GW
Tel.: 01309 671111
www.jamesjones.co.uk
Email: jji-joists@jamesjones.co.uk

For details of "JES" Joist End Support:
Simpson Strong Tie,
Winchester Road, Cardinal Point, Tamworth, Staffordshire, B78 3HG
Tel.: 01827 255600
www.strongtie.co.uk
Email: uktechnical@strongtie.com

Robinson Manufacturing Ltd,
Units 25–31, Meadow Close, Ise Valley Industrial Estate, Wellingborough, Northants, NN8 4BH
Tel.: 01933 277701
www.robinsonmanufacturing.co.uk
Email: sales@robinsonmanufacturing.co.uk

BUILDING REGULATIONS
'Guide to Building Control', by Anthony Gwynne. Published by Wiley-Blackwell. ISBN 978-0-470-65753-9.

BUILDING RESEARCH ESTABLISHMENT PUBLICATIONS AND BRITISH STANDARDS
All of the publications listed in this section below can be obtained from:
BRE Bookshop, Garston, Watford, WD25 9XX
Tel.: 01923 664761
www.brebookshop.com
or
IHS Rapidoc (BRE Bookshop) Willoughby Road, Bracknell, RG12 8DW
Tel.: 01344 404407
Email: brebookshop@ihsrapidoc.com

BS 8000-6:2013, Workmanship on Building Sites. Code of Practice for Slating and Tiling of Roofs and Walls.

BS 5534:2014, Slating and Tiling for Pitched Roofs and Vertical Cladding – Code of Practice.

BS 5534:2003+A1:2010, Code of Practice for Slating and Tiling (including Shingles).

NHBC Technical Standards, Part 7 – Roofs. Chapter 7.2 Pitched Roofs.

BS 6399-2:1997, Loading for Buildings. Code of Practice for Wind Loads.

BS EN 13859-1:2014, Flexible Sheets for Waterproofing. Definitions and Characteristics of Underlays. Underlays for Discontinuous Roofing.

BRE Digest DG489 Wind Loads on Roof-Mounted Photovoltaic and Solar Thermal Systems.

BRE Digest DG495 Mechanical Installation of Roof-Mounted Photovoltaic Systems.

DESIGN & CONSTRUCTION
Mindham, C. (2006) *Roof Construction & Loft Conversion*, 4th edn. 256 pp. Blackwell Publishing, Oxford. ISBN 1-4051-3963-3. Available through bookshops.

National House-Building Council (NHBC), NHBC House, Davy Avenue, Milton Keynes, Bucks MK5 8FP
Tel.: 0844 633 1000
www.nhbc.co.uk
2016 NHBC Standards Book.

Span Tables for Solid Timber Members in Floors, Ceilings and Roofs (Excluding Trussed Rafter Roofs) for Dwellings. Technical publication published by TRADA Technology Design Aid, DA1/2004, ISBN 1900510464. Available from TRADA Technology, Chiltern House, Stocking Lane, Hughenden Valley, High Wycombe, Bucks, HP14 4ND.
   Tel.: 01494 569600
   www.trada.co.uk
   Email: information@trada.co.uk

Eurocode 5 span tables for solid timber members in floors, ceiling and roofs for dwellings 4th Edition ISBN 978-1-909594-14-2 available from Trada Technology as above.
   Trussed Rafter Association
   The Building Centre, 26 Store Street, London, WC1E 7BT
   Tel.: 020 320 0032
   www.tra.org.uk
   Email: info@tra.org.uk

DRY ROOFING SYSTEMS: Ridge, Valley, Verge and Hip
   Monier Redland Ltd,
   Manor Business Park, Gatwick Road, Crawley, West Sussex, RH10 9NZ
   Customer Service Hot Line Tel.: 0870 560 1000
   www.redland.co.uk
   Email: sales.redland@monier.com

   PERMAVENT Ltd,
   11 Cumberland Drive, Granby Industrial Estate, Weymouth, DT4 9TB
   Tel.: 01305 766703

www.permavent.co.uk
Email: info@easyroofd.co.uk

Marley Eternit Ltd,
Lichfield Road, Branston, Burton on Trent, DE14 3HD
Customer Service Tel.: 01283 722588
www.marleyeternit.co.uk
Email: info@marleyeternit.co.uk

INSULATION

Boulder Developments Ltd,
BHF, Norwell, Nottinghamshire, NG23 6JN
Tel.: 01636 639900
www.superfoil.co.uk
Email: sales@bhunlimited.co.uk

Kingspan Insulation Ltd,
Pembridge, Leominster, Herefordshire, HR6 9LA
Tel.: 01544 388601
www.kingspantek.co.uk
Email: info@kingspaninsulation.co.uk

INTEGRATED SOLAR PANELS AND FIXINGS

Sandtoft Roof Tiles Ltd,
Brooks Drive, Cheadle Royal Business Park, Cheadle, Cheshire, SK8 3SA
Tel.: 0844 9395 900

www.sandtoft.com
Email: info@sandtoft.co.uk
Ref: PV48 System

Monier Redland Ltd,
Manor Business Park, Gatwick Road, Crawley, West Sussex, RH10 9NZ
Customer Service Hot Line Tel.: 0870 560 1000
www.redland.co.uk
Email: sales.redland@monier.com
Ref: 250 PV System

Klober Ltd,
Unit 6F, East Midlands Distribution Centre, Short Lane, Castle Donnington,
Derbyshire, DE74 2HA
Tel.: 0845 600 4427
www.roof-flashing.info
Email: info@haus-profi.co.uk

See also under 'Building Research Establishment Publications'.

See text in Chapter 11 for useful websites.

LEAD SUBSTITUTES

Klober Ltd,
Unit 6F, East Midlands Distribution Centre, Short Lane, Castle Donnington,

Derbyshire, DE74 2HA
Tel.: 0845 600 4427
www.roof-flashing.info
Email: info@haus-profi.co.uk

Ubbink (UK) Ltd,
33 Liliput Road, Brackmills, Northampton, NN4 7DT
Tel.: 01604 433000
www.ubbink.co.uk
Email: info@ubbink.co.uk

METAL ROOF TILES
Metrotile UK Ltd
Unit 3, Sheldon Business Park, Sheldon Corner,
Chippenham, Wiltshire, SN14 0RQ
Tel: 01249 658514
www.metrotile.co.uk
Email: info@metrotile.co.uk

Britmet Tileform Ltd
Spital Farm Offices, Thorpe Mead
Banbury, Oxfordshire, OX16 4RZ
Tel: 01295 250998
www.britmet.co.uk
Email: sales@britmet.co.uk

ROOFING METALWORK

Simpson Strong Tie,
Winchester Road, Cardinal Point, Tamworth, Staffordshire, B78 3HG
Tel.: 01827 255600
www.strongtie.co.uk
Email: uktechnical@strongtie.com

ROOFING PUTTIES

Building Products Ltd,
12 Wells Promenade, Ilkley, West Yorkshire, LS29 9LF
Tel.: 01943 607538
www.fleximroofputty.com
Email: info@buildingproducts.co.uk

Monier Redland Ltd,
Manor Business Park, Gatwick Road, Crawley, West Sussex, RH10 9NZ
Customer Service Hot Line Tel.: 0870 560 1000
www.redland.co.uk
Email: sales.redland@monier.com
The Redland Pocket Guide

ROOF TILES AND SLATES

Monier Redland Ltd,
Manor Business Park, Gatwick Road, Crawley, West Sussex, RH10 9NZ
Customer Service Hot Line Tel.: 0870 560 1000

www.redland.co.uk
Email: sales.redland@monier.com
The Redland Pocket Guide

Marley Eternit Ltd,
Lichfield Road, Branston, Burton on Trent, DE14 3HD
Customer Service Tel.: 01283 722588
www.marlcyeternit.co.uk
Email: info@marleyeternit.co.uk
Roofing Sitework Guide

ROOF TILE AND SLATE FIXING
Monier Redland Ltd,
Manor Business Park, Gatwick Road, Crawley, West Sussex, RH10 9NZ
Customer Service Hot Line Tel.: 0870 560 1000
www.redland.co.uk
Email: sales.redland@monier.com
The Redland Pocket Guide

Marley Eternit Ltd,
Lichfield Road, Branston, Burton on Trent, DE14 3HD
Customer Service Tel.: 01283 722588
www.marleyeternit.co.uk
Email: info@marleyeternit.co.uk
Roofing Sitework Guide

ROOF VENTILATION

Monier Redland Ltd,
Manor Business Park, Gatwick Road, Crawley, West Sussex, RH10 9NZ
Customer Service Hot Line Tel.: 0870 560 1000
www.redland.co.uk
Email: sales.redland@monier.com
The Redland Pocket Guide

Marley Eternit Ltd,
Lichfield Road, Branston, Burton on Trent, DE14 3HD
Customer Service Tel.: 01283 722588
www.marleyeternit.co.uk
Email: info@marleyeternit.co.uk

PERMAVENT Ltd,
11 Cumberland Drive, Granby Industrial Estate, Weymouth, DT4 9TB
Tel.: 01305 766703
www.permavent.co.uk
Email: info@easyroofd.co.uk

STRUCTURAL ROOFING PANEL/SPANDREL PANELS

Kingspan Insulation Ltd,
Pembridge, Leominster, Herefordshire, HR6 9LA
Tel.: 01544 388601
www.kingspantek.co.uk
Email: info@kingspaninsulation.co.uk

Donaldson Timber Engineering,
James Donaldson & Sons Ltd, Suite A, Haig House, Haig Business Centre, Balgonie Road,
Markinch KY7 6AQ
Tel.: 01592 752244
www.donaldsontimber.com
Email: dteb@donaldson-timber.co.uk

James Jones and Sons Ltd,
Timber Systems Division, Greshop Industrial Estate, Forres, Scotland, IV36 2GW
Tel.: 01309 671111
www.jamesjones.co.uk
Email: jji-joists@jamesjones.co.uk

Robinson Manufacturing Ltd,
Units 25–31, Meadow Close, Ise Valley Industrial Estate, Wellingborough, Northants, NN8 4BH
Tel.: 01933 277701
www.robinsonmanufacturing.co.uk
Email: sales@robinsonmanufacturing.co.uk

TIMBER 'I' BEAMS/JOISTS

James Jones and Sons Ltd,
Timber Systems Division, Greshop Industrial Estate, Forres, Scotland, IV36 2GW
Tel.: 01309 671111
www.jamesjones.co.uk
Email: jji-joists@jamesjones.co.uk

Donaldson Timber Engineering,
James Donaldson & Sons Ltd, Suite A, Haig House, Haig Business Centre, Balgonie Road,
Markinch KY7 6AQ
Tel.: 01592 752244
www.donaldsontimber.com
Email: dteb@donaldson-timber.co.uk

Trussed Rafter Association
The Building Centre, 26 Store Street, London, WC1E 7BT
Tel.: 020 320 0032
www.tra.org.uk
Email: info@tra.org.uk

For list of manufacturers.

Robinson Manufacturing Ltd,
Units 25–31, Meadow Close, Ise Valley Industrial Estate, Wellingborough, Northants, NN8 4BH
Tel.: 01933 277701
www.robinsonmanufacturing.co.uk
Email: sales@robinsonmanufacturing.co.uk

UNDERLAYS
PERMAVENT Ltd,
11 Cumberland Drive, Granby Industrial Estate, Weymouth, DT4 9TB
Tel.: 01305 766703

www.permavent.co.uk
Email: info@easyroofd.co.uk

Monier Redland Ltd,
Manor Business Park, Gatwick Road, Crawley, West Sussex, RH10 9NZ
Customer Service Hot Line Tel.: 0870 560 1000
www.redland.co.uk
Email: sales.redland@monier.com

Boulder Developments Ltd,
BHF, Norwell, Nottinghamshire, NG23 6JN
Tel.: 01636 639900
www.superfoil.co.uk
Email: sales@bhunlimited.co.uk

The Proctor Group
The Haugh, Blairgowrie
Perthshire, PH10 7ER
Tel: 01250 872261
www.proctorgroup.com
Email: contact@proctorgroup.com

INDEX

Goss's Roofing Ready Reckoner: From Timberwork to Tiles, Fifth Edition. C. N. Mindham.
© 2016 John Wiley & Sons, Ltd. Published 2016 by John Wiley & Sons, Ltd.